POLITICS, L
ISLAM I

This book offers a unique insight into the changing nature of power and hierarchy in rural Pakistan from colonial times to present day. This book shows how electoral politics and the erosion of traditional patron–client ties have not empowered the lower classes. The monograph highlights the persistence of debt bondage and illustrates how electoral politics provides assertive landlord politicians with opportunities to further consolidate their power and wealth at the expense of subordinate classes. It also critically examines the relationship between local forms of Islam and landed power.

The volume will be of interest to scholars and researchers of Pakistan and South Asian politics, sociology, social anthropology and Islam, as well as economics, development studies and security studies.

Nicolas Martin is Senior Research Associate at University College London, UK.

EXPLORING THE POLITICAL IN SOUTH ASIA
Series Editor: Mukulika Banerjee
*Associate Professor, Department of Anthropology,
London School of Economics and Political Science*

Exploring the Political in South Asia is devoted to the publication of research on the political cultures of the region. The books in this series present qualitative and quantitative analyses grounded in field research, and explore the cultures of democracies in their everyday local settings, specifically the workings of modern political institutions, practices of political mobilisation, manoeuvres of high politics, structures of popular beliefs, content of political ideologies and styles of political leadership, among others. Through fine-grained descriptions of particular settings in South Asia, the studies presented in this series inform, and have implications for, general discussions of democracy and politics elsewhere in the world.

Also in this series

THE VERNACULARISATION OF DEMOCRACY
Politics, Caste and Religion in India
Lucia Michelutti
978-0-415-46732-2

RISE OF THE PLEBEIANS?
The Changing Face of the Indian Legislative Assemblies
Edited by Christophe Jaffrelot and Sanjay Kumar
978-0-415-46092-7

BROADENING AND DEEPENING DEMOCRACY
Political Innovation in Karnataka
E. Raghavan and James Manor
978-0-415-54454-2

RETRO-MODERN INDIA
Forging the Low-caste Self
Manuela Ciotti
978-0-415-56311-6

POWER AND INFLUENCE IN INDIA
Bosses, Lords and Captains
Edited by Pamela Price and Arild Engelsen Ruud
978-0-415-58595-8

DALITS IN NEOLIBERAL INDIA
Mobility or Marginalisation?
Edited by Clarinda Still
978-1-13-802024-5

WHY INDIA VOTES?
Mukulika Banerjee
978-1-13-801971-3

CRIMINAL CAPITAL
Violence, Corruption and Class in Industrial India
Andrew Sanchez
978-1-138-92196-2

THE POLITICS OF CASTE IN WEST BENGAL
Edited by Uday Chandra, Geir Heierstad, Kenneth Bo Nielsen
978-1-13-892148-1

POLITICS, LANDLORDS AND ISLAM IN PAKISTAN

Nicolas Martin

LONDON AND NEW YORK

First published 2016 by Routledge
2 Park Square, Milton Park, Abingdon, Oxfordshire OX14 4RN
711 Third Avenue, New York, NY 10017

Routledge is an imprint of the Taylor & Francis Group, an informa business

First issued in paperback 2017

Copyright © 2016 Nicolas Martin

The right of Nicolas Martin to be identified as author of this work has been asserted in accordance with sections 77 and 78 of the Copyright, Designs and Patents Act 1988.

All rights reserved. No part of this book may be reprinted or reproduced or utilised in any form or by any electronic, mechanical, or other means, now known or hereafter invented, including photocopying and recording, or in any information storage or retrieval system, without permission in writing from the publishers.

Notice:
Product or corporate names may be trademarks or registered trademarks, and are used only for identification and explanation without intent to infringe.

British Library Cataloguing-in-Publication Data
A catalogue record for this book is available from the British Library

Library of Congress Cataloging-in-Publication Data
A catalog record has been requested for this book

ISBN: 978-1-138-82188-0 (hbk)
ISBN: 978-0-8153-9298-9 (pbk)

Typeset in Goudy
by Apex CoVantage, LLC

CONTENTS

Foreword by Mukulika Banerjee		viii
Acknowledgements		xi
Note on transliteration		xiii
	Introduction	1
1	The village	20
2	Debt and bondage	44
3	Electoral politics and the reproduction of inequality	66
4	The enemy of my enemy is my friend	93
5	Elections and devolution	120
6	Islam, selflessness and prosperity	145
	Conclusion	168
	Bibliography	177
	Glossary	186
	Index	189

FOREWORD

When we set up the series 'Exploring the Political in South Asia', we had intended that we would be able to include within it rich studies of the political in all the countries of the Indian subcontinent, not just India on which the literature is without doubt the richest. It is therefore a particular pleasure to write the foreword to this volume as it is the first book on Pakistan that we are publishing in this series. I am delighted to note that Nicolas Martin has delivered a book that is unique and in keeping with the tradition of rigorous and insightful political ethnography that this series attempts to showcase. Martin presents an analysis of contemporary Pakistan on the basis of evidence gathered during a period of two years during which he lived in rural Punjab – sometimes in a Sufi shrine, sometimes in humble rural homes, made friends with locals from across classes, worked in the local languages and got to know their public and private lives with an intimacy only anthropologists dare to achieve. This book therefore adds a totally new perspective and insight into Pakistan, a country that is much written about by writers who gain but a fleeting familiarity with people and places and usually from a distance. Martin's account adds to this corpus with a scholarly and evidence-based account of the Pakistani state, electoral politics and clientelism – written with a nose firmly on the ground.

This book tackles the most knotty issue that lies at the heart of politics in the populous province of Punjab – that of patronage. Almost any account of Pakistani politics and its chequered record of democracy has drawn attention to the importance of political patronage in the creation of vote blocs that are mobilised for elections. These accounts have drawn attention to the entrenched nature of the power of rural elites who keep an impoverished rural workforce under their thumb by mediating between the state's resources and the intended recipients. Such clientelistic relationships are then mobilised for votes during the election season as clients are forced to support their patrons' political loyalties to ensure their own survival. Scholars in their attempt to interpret such a situation empathetically have sometimes argued that exploitative as it might seem from the outside, such a system of patronage has its advantages as it alleviates poverty by at least making sure that state resources are allocated somewhere and in doing so possibly also stem the possibility of widespread protest movements among the rural poor.

Martin categorically rejects such a reading of rural Pakistan. His book makes it clear that patronage is nothing but 'locally embedded despotism'. He shows how landlords continue to use debt bondage as a way of controlling their clients and the other ways in which elites continuously exploit institutional, religious and social mechanisms to reproduce their power. He shows how the combination of feudal patronage and electoral democracy undermines other social mechanisms of creating social solidarity through political parties representing sectoral interests. As a result all alliances are forced to become nakedly instrumental and transactional in the zero sum game of political power. The result is everyday violence and vendetta that in turn create a state of vulnerability and fear among the poor. In the long run, this kind of politics, therefore, undermines the everyday power of the state by stemming the effectiveness of public sector services and benefits that the state is able to provide, by capturing resources for personal gain. Martin conducted this study while a military regime was in power under President Musharraf. Many Pakistanis had hoped that the strong leadership at the centre would make such local factional politics irrelevant as a military government would undercut corrupt political structures to reach the people directly. But unfortunately, even military regimes need to create a legitimacy for their existence, and Martin notes, 'military regimes used devolution programmes to create a loyal class of politicians'.

Unsurprisingly, the ordinary members of the chronically poor, rural agricultural workforce struggle to provide a robust challenge to such blatant corruption of the political establishment. The workforce is already divided by caste membership, and rising Islamism diverted attention from issues of inequality and social justice. Further, Martin points out, military meddling made rural politics more parochial and kinship based than ever before and brought personal enmities to the fore. As a result, instead of providing a radical critique of an exploitative system of extraction and abuse, and the system that causes their own structural subordination, ordinary people came up with only moral critiques of individual elites and their behaviour, judging one against the other.

But Nicolas Martin is able to end on a more upbeat note by bringing his analyses up to the present. He does this by evaluating the implications of the recent run of democratic governments that Pakistan has seen, first with the full term of the Pakistan People's Party government and then the relatively peaceful electoral change that brought their old rivals, the

FOREWORD

Pakistan Muslim League (Nawaz) back into power. The PPP government during its tenure under Zardari's leadership was able to bring in legislative changes that affected all three arms of power of a democratic government: it strengthened executive governance from political interference; it limited the ability of the military leadership to destabilise elected governments; and it reduced the executive's role in appointing members of the judiciary. Crucially, the changes also made Pakistan's Electoral Commission more accountable in its conduct by making its processes more transparent and subject to scrutiny. All of these measures could potentially have a far-reaching impact on Pakistani politics. With growing urbanisation and cleaner governance, we can only hope that Pakistan's despotic demagogic rural landlords will soon be consigned to history.

Mukulika Banerjee
London
July 2014

ACKNOWLEDGEMENTS

I would first of all like to thank the European Research Council and the UK Economic and Social Research Council (ERC-2011-StG – N° 284080— AISMA and the UK Economic and Social Research Council) for funding me during the year 2012–13 writing this book. I would additionally like to thank a number of people for their generous support in both Pakistan and the UK. I am particularly indebted to my supervisors, Professor Parry and Professor Mundy, for their patient guidance and careful reading of my work. Their questions and suggestions always pointed me in more productive directions. Dr Stafford, a supervisor through my initial months as a PhD candidate, provided me with equally helpful advice. I would also like to thank Professor Deborah James, Dr Alpa Shah, Dr Amita Baviskar, Dr Mukulika Banerjee, Dr Jens Lerche, Dr Magnus Marsden, Dr Lucia Michelutti, Dr Tom Bolyston, Dr Anastasia Piliavski, Dr Maxim Bolt, and Dr Andrew Sanchez for reading and commenting on different chapters in this book. Likewise the comments of the anonymous reviewer at Routledge helped me clarify certain arguments. Finally I would like to thank Professors Jan Breman and Stuart Corbridge for their suggestions on how to turn my thesis into this book.

I would also like to thank the people who commented on papers I presented at different seminars in UK universities, as well as the various anonymous reviewers who read and gave feedback on papers I sent out for publication. More informally, my discussions on Pakistani politics with Dr Hassan Javid from the LSE were particularly enlightening.

Dr Anwar Iqbal and Professor Hafeez-ur Rahman at Quaid-I Azam University in Islamabad were most kind in supporting my affiliation and in offering me their hospitality, guidance and help during my stay in Pakistan. Dr Anwar Iqbal in particular went far beyond the call of duty by spending several days in government departments trying to get my research visa approved.

ACKNOWLEDGEMENTS

All of my fellow students in the LSE Department of Anthropology and elsewhere were very supportive, and I benefited from the frequent exchange of ideas about our research topics and everything else. I am particularly indebted to Roland de Wilde who remained in contact with me throughout fieldwork and who subsequently read drafts of several of my chapters. Among others I would also like to thank Carrie Heitmeyer, Amit Desai, Mette High, Vicky Boydell, Giovanni Bocchi, and Pradeep Shinde for discussing my work with me. Last but not least I am deeply indebted to Jonathan Grossman for long hours spent meticulously reading my thesis and giving me extensive editorial comments.

For offering me his hospitality and friendship while I was in Islamabad I am grateful to Anibal Oprandi from Plan International. He always made me feel welcome at his house, and his enthusiastic support for my research project encouraged me to persevere. I am also grateful to Rodolfo and Susana Martin-Saravia for their wonderful hospitality in Islamabad. I would also like to thank Khaled Khan from Charsadda who introduced me to people from all walks of life in Islamabad and who helped me overcome significant bureaucratic hurdles. To Abdul Aziz Abassi I am indebted for having ignited my interest in Pakistan.

In Bek Sagrana – the pseudonym used to refer to my village fieldsite – I am indebted to a large number of people who would prefer to remain unnamed. They were generous and hospitable and willing to discuss my research with me. Without their friendship and support during difficult times it would have been impossible for me to complete this book.

Finally I would like to thank my wife Emily for her affectionate support and encouragement and for tolerating my obsessive focus on all things related to Pakistan during the past several years.

NOTE ON TRANSLITERATION

The first instance of all Urdu and Punjabi words is italicised, and thereafter italics are dropped. For the sake of simplicity, the plural of Urdu and Punjabi words is denoted by the addition of an 's'. English translations have been used where they were deemed adequate to convey the meaning of the original.

The exchange rate at the time of fieldwork was roughly Rs100 for £1. The name of the village, the names of villagers and the name of the dominant caste within it have all been changed in order to protect the identity of informants.

INTRODUCTION

In his collection of short stories entitled *In Other Rooms, Other Wonders*, Daniyal Mueenuddin describes a feudal world reminiscent of that described by 19th-century Russian novelists that inspire his work. It is a world inhabited by absentee landlords with vast estates in the Southern Punjab, large retinues of servants, peasants, cattle rustlers, criminals, peasant girl mistresses and unscrupulous farm managers, bureaucrats, and politicians. Like their counterparts in 19th-century Russia, these cosmopolitan Punjabi feudal lords admire the cultural achievements of Europe where they spend their summers. Their mansions in Lahore are staffed with valets, butlers, drivers, cooks, and maids, all brought from the villages on their estates. Here parties are hosted where high-ranking military officers, politicians, businessmen, and civil servants gather and drink smuggled Black Label Scotch whisky. Their children study in prestigious universities in the United Kingdom and, increasingly, the United States.

However, the world that Mueenuddin describes is changing. Members of this landed aristocracy are no longer necessarily the wealthiest or the most politically influential people in the country. The wealthiest aristocratic (*ashrafi*)[1] landlords through whom the British once used to rule the Punjab, and who remained politically prominent after independence, have been gradually overtaken both politically and economically by prosperous Jat farmers as well as by businessmen. In rural Pakistan, the landlord class continues to wield disproportionate power, but there have been some power shifts within it. As Mueenuddin's short stories and this work document, those anglicised aristocratic feudal lords who fail to invest in agricultural modernisation, and who shy away from the feuds and from the violence of

---

[1] The term *ashrafi* in South Asia generally refers to people of noble descent. Amongst the Muslims the ashrafi landlords were predominantly those claiming Arabian and Central Asian descent.

rural Punjabi politics, are overtaken politically and economically by traditionally less wealthy and influential members of the landed classes. The aristocrats claim that the country is now run by corrupt and uneducated upstarts who flaunt their wealth through lavish weddings and garishly decorated mansions.

In Mueenuddin's short stories the decline of the aristocratic order is illustrated through the case of the absentee landlord K.K. Harouni. Harouni, who lives in a large air-conditioned mansion in Lahore and who only occasionally visits his vast estate, is oblivious to the fact that he is being swindled by the manager, a local landlord, to whom he has delegated the estate's management. When K.K. Harouni needs money to invest in unsuccessful business ventures – in order to keep up with upstart industrialists and to plug holes into the leaky finances of his Lahore mansion – he trustingly gives his manager Jaglani powers of attorney to sell some of his land. Since Harouni doesn't even know the value of his lands, Jaglani buys the plots himself at prices far below market value and soon owns several hundred acres. Jaglani runs the estate with an iron fist, more for his own benefit than for his master's. To villagers Harouni is a distant figure, and they exploit his careless attitude towards his finances by stealing from him whenever they can. While Harouni leads a cosmopolitan lifestyle in Lahore and abroad, it is Jaglani who deals with the grittier aspects of life on the estate and in the surrounding area. He supervises labourers, deals with village disputes and works with local toughs to achieve control over the area's inhabitants. Eventually Jaglani's local influence allows him to become a provincial minister. Although Mueenuddin does not say so, political office probably allows him to further consolidate his landholdings through the multiple opportunities for enrichment it offers. The present work is interested in landlords that resemble Jaglani more closely than they do K.K. Harouni.

Anatol Lieven writes: 'I thought of arguing that there is no such thing as a feudal in Pakistan; but then I remembered wild-boar hunting with the noble landowners of Sindh – a remarkably feudal experience' (Lieven 2011: 17). I too witnessed tent-pegging and hunting events that had a feudal feel to them. At the Lahore horse and cattle show in 2004, landlords wearing starched white shalwar kameezs displayed their martial prowess on frisky stallions. Between events they lounged on charpais surrounded by servants attending to their every need. In the evenings they watched bangled horses dancing, and later on in the night they watched dancing girls in Lahore's red light district or in more exclusive private settings. One evening a landlord boasted that he could spend the night with any girl in his home village that caught his fancy, regardless of their husband or father's objections. I later learnt that this was no idle boast.

This event at the start of my fieldwork and my initial forays into the Punjabi countryside – where I soon became aware of the arbitrary despotism of landlords and of the widespread practice of debt-bondage – seemed to confirm the feudal hypothesis. The landlords I saw appeared to be diverting much of their surplus into non-productive investments in maintaining local power, ranging from consumption loans to collective social rituals to the hire of armed thugs (*goondas*). Moreover they spent most of their time socialising with other landlords rather than managing their farms to maximise agricultural productivity. The impression that they were using both their capital and their time unproductively was reinforced by academic and policy literature according to which semi-feudal power structures in South Asian villages simultaneously exploit the poor and depress economic growth potential (cf. Harris 1980).

Particularly during the two decades after independence, the policy literature in South Asia advocated land reforms on the basis that large farms were inefficient because they were less labour intensive than smaller ones. However somewhat later literature (Alavi 1973, Khan and Maki 1975, Herring 1983) indicated that as modern technology and production relations penetrated South Asian agriculture, smaller farms without access to improved capital works, technical information and inexpensive credit lost their traditional yield advantage. According to Herring (1983), local elites invested in supra-local politics in order to get access to these and to bolster labour repressive social organisations of production which traditional patron–client ties no longer sustained. In his view large farmers in Pakistan were capitalists, and this was entirely compatible with labour repressive practices such as debt-bondage. While Ashok Rudra (1980) viewed landlords' investment in maintaining local power – through consumption loans, collective social rituals and the hire of goondas – as non-productive, Herring argued that these could be considered as functional requisites for the reproduction of labour comparable to expenditure on antitrust lawyers, industrial spies, plant security guards, public relations personnel, and corporate philanthropy under industrial capitalism.

When I started fieldwork the discussion that followed from the comment of a powerful landlord and hereditary saint (*pir*) seemed to indicate that landlords – like Mueenuddin's Jaglani – could no longer take their power for granted in the face of both capitalist development and electoral politics.

The *pir* came from the Punjabi district of Jhang and turned up at our tent at the Horse and Cattle Show near the stables of Lahore's Forter Stadium accompanied by four servants, one of which was carrying a hooded hawk on a gloved hand. He sat on a charpai and, between puffs on an ornate brass hukka with a revolving base, embarked on a monologue lamenting the growing political and economic clout of upstart businessmen. He claimed that a wealthy industrialist had expressed envy because he had so many servants; according to him this showed how money could never buy the social influence that he enjoyed as a landlord. However his comment that the introduction of schools decreased the labour force at his disposal – by creating unrealistic ambitions among workers – hinted at the fact that he could no longer take this social influence for granted. As soon as he left, one of the assembled landlords turned to me with a mischievous look and asked me what I thought about Pakistan's feudal culture, at which the other landlords burst into laughter. One of them went on to tell me that the *pir* was anachronistic and trying to revive past glories and that in modern day Pakistan you needed to get involved in both politics and business if you wanted to retain your social influence and status. He said that modern landlords needed to make money by investing in their farms, into businesses, into education, and into politics.

The principal focus of this book is on how in contemporary Pakistan control over the state apparatus is central to landlords' strategies of accumulation and domination. In it, I show how members of a numerically strong patrilineal clan (*biraderi*)[2] of prosperous Jats consolidated their political and economic power thanks to their cohesion and to their forceful involvement in local politics and in frequently criminal business ventures. In this book, as in other works I have published, I have changed my informants' names to protect their identities. I have also changed the name of their subcaste to Gondal for the same reason.[3] Like Jaglani in Mueenuddin's short story, these Gondal Jats overtook members of an aristocratic lineage of landlords, often more concerned with polo and hare-coursing than politics and business, as the political overlords of a region of the Sargodha District.

Here it should be noted that while the image of a decaying aristocracy is a powerful one, it doesn't fully portray reality. Unlike the Jaglanis and the Gondals of this world, the K.K. Harounis and Makhdooms could afford to move out of the local political scene. Having degrees from prestigious

---

[2] In the Pakistani Punjab the terms 'biraderi' and 'qaum' are often used interchangeably to refer to the patrilineal clan although the latter is more frequently used for more inclusive forms of identity such as ethnicity (e.g. Pathan, Baluch, Punjabi) or nationality (Pakistani, Indian, etc.). The term caste (*zat*) is not very frequently used. See especially Chaudhary (1999: 11).

[3] Needless to say the events and characters in this book have nothing to do with members of the actual Gondal subcaste. I have chosen the name Gondal merely because it is large sub-caste of Jats in district Sargodha and is of roughly equivalent rank to that of my informants.

universities in Pakistan and abroad – mainly in the United Kingdom and the United States – they could pursue lucrative careers in banks and multinational corporations and therefore avoid the cut-throat world of contemporary Pakistani politics. Others from this class obtained high ranking posts in the bureaucracy and then frequently moved into international organisations such as the United Nations, the World Bank and the Asian Development Bank. Those with the most foresight invested in industry, building sugar mills and food processing plants. Although by no means all aristocratic landlords in Pakistan opted out of rural politics, anecdotal evidence suggests that many followed a similar path to the Makhdooms. In the area of Bek Sagrana, the fictional name I have given to my village, the effect of their exit was to open the door to the forceful and numerous members of the Gondal clan.

The case of Chowdri Sahib – as I shall call him here – the heir of a powerful aristocratic family in Sargodha district nicely illustrates why aristocratic landlords withdrew from local politics, as well as what happened to their constituencies when they did. Chowdri Sahib became a member of the provincial assembly in his early thirties. His family's vast estate and agro-industries employing tens of thousands of people gave him a virtually guaranteed seat in the national assembly regardless of his age and lack of experience. Brought up in Lahore, and later at boarding schools abroad, he hadn't found rural Punjabi politics to his taste and decided that there was no need for him to remain involved. He complained about how his constituents had incessantly assailed him with requests to remove uncooperative police officers and local land registration officers (*patwaris*) in order to pursue some land grab (*qabza*) or other illicit scheme. Several landlords apparently asked him to get their servants government jobs so that the state would pay their wages while they continued to work for them. People also frequently approached him to arrange release from jail for relatives guilty of some crime or other.

After abandoning politics Chowdri Sahib spent less time in the village and his absence created a power vacuum that allowed four forceful brothers – medium-sized landholders – to start harassing people and grabbing their land. Responding to the complaints of villagers, Chowdri Sahib eventually took the four siblings to court. The brothers responded with threats of violence, and Chowdri Sahib and his family had to start travelling to the village with armed guards. This situation lasted for four years, during which time several people were killed and injured in gun battles. The eldest of the four brothers was himself killed by Chowdri Sahib's gunmen after he attacked them when they were driving through the village.

The remaining three siblings later launched a second attack on Chowdri Sahib's gunmen when these were drinking tea at a roadside stall and accidentally shot a bystander in the spine. Immediately after the shootout, a lawyer – a man with lots of experience in rural litigation – instructed Chowdri Sahib to quickly fabricate evidence to prove that his men had been attacked first. He told him to do what was usually done in such circumstances, namely to get one of his gunmen to shoot himself in an arm or a leg and then go to the police station to register a First Information Report (FIR).[4] Chowdri Sahib, who had little experience with this type of conflict and litigation, thought the suggestion was absurd and it was only when he was warned that his opponents would do it first if he didn't that he finally ordered one of his gunmen to shoot himself in the arm.

While these practices came naturally to sons of the soil types like the landlords who are the subject of this book, they didn't to Chowdri Sahib. He personally told me that the easiest thing would have been to act in true 'feudal' fashion and simply got the troublemakers killed in a fake encounter. A high ranking Inspector of Police even allegedly advised him to do this, and some senior judges pledged to help him sort the case out. But Chowdri Sahib understandably wanted to do things through the courts with as little violence as possible, so it took him four years to sort out a situation that he could have sorted out within days. The landlords who are the subject of this book didn't shy away from corruption, crime and violence as Chowdri Sahib did, but their more urbanised and educated children may one day do so and decide to opt out of local politics.

The ethnography presented in this book looks at how electoral politics led to the emergence of a new class of forceful political entrepreneurs from within the ranks of the middle and upper-middle sections of the landed class, and to the partial displacement of the traditional aristocratic landed elite.[5] Authoritarian military regimes buttressed this class – often by undermining the impartiality of the state bureaucracy and the judiciary on their behalf – in order to maintain a political support base. As Hassan Javid (2011) has shown, the landed elite remained powerful in Pakistan because authoritarian regimes used them to buttress their rule, much like the British colonial state had done previously. Thus although – as Hamza Alavi (1972b) once argued – the Pakistani postcolonial state has acted as a mediator between the often contradictory interests of capitalist and feudal factions within the ruling class, its relationship with 'the landed class grew

---

[4] Another common practice in these circumstances was to get supporters and gunmen to break a finger or an arm.

[5] As in Chakravarti's account of Rajasthani politics control over the land was insufficient to guarantee political power in the face of 'political entrepreneurs' (Chakravarti 1975: 210).

to become one of mutual structural dependence' (Javid 2011: 358). From the anthropological worm's eye view, this book illustrates how this mutual structural dependence between the postcolonial state and the elite plays out on the ground and affects ordinary people's lives.

This would indicate that while the local political influence of the landed aristocracy has been dented the dominance of the landed class more broadly has not. This book explores whether political and economic change has in any way diminished the power of landed elites.

The book also questions Lieven (2011) and Lyon's (2005) claim that landlords in Pakistan make life more tolerable for the poor because without them ordinary villagers would find it far more difficult to get access to things like shelter, cash loans and justice, or even to state resources such as health care. Lieven goes further and claims that because such patronage mitigates poverty and exploitation, it helps prevent large scale popular revolt. From the tradition of subaltern studies, and with a focus on India, Partha Chatterjee (2004, 2011) makes a similar argument, claiming that politicians who broker state resources – often by circumventing the law and bureaucratic procedure – enrich democracy by bringing the unresponsive state institutions closer to people. While these authors come from different intellectual traditions, they both effectively agree that patronage – or clientelism – are at worst lesser evils. This book is less positive about clientelistic politics, and its argument is closer to Nelson's who claims that what Chatterjee describes 'is not a democracy, but a particular form of locally embedded despotism' (Nelson 2011: 230).[6] However, while Nelson's book about rural politics in the Punjab focuses specifically on political mediation in land disputes, this book examines landed power more broadly. Nelson's work focuses on political power, the law and kinship ties; this book examines the broader structural sources of political power. Thus this book focuses on class and ideology as well as kinship, and is therefore closer to earlier sociological work by authors like Hamza Alavi and Saghir Ahmed than to Matthew Nelson's work. Above all this is ethnography of a system of domination in the rural Pakistani Punjab.

Fieldwork

My entry into Bek Sagrana was made possible through Pakistani contacts and friends at the London School of Economics. I sought their advice as to how I might be able to carry out research on what Pakistanis referred to as

---

[6] I also elaborate on this argument in a paper entitled 'The Dark Side of Political Society: Patronage and the Reproduction of Social Inequality in the Pakistani Punjab' (Martin 2014).

'feudalism', and they put me in contact with the Gondals of Bek Sagrana whom they knew to be extensively involved in politics at the provincial and national levels. The man who first welcomed me to the village was Mahmood Abbas Gondal.[7] He was the youngest son of the pious Sufi Ahmed Abbas Gondal, and a member of one of the wealthiest households in the village. Mahmood Abbas Gondal spent most of his time in Lahore where he had succeeded in securing places for his two sons at Aitchison College. He was a lawyer by training but had opted not to pursue a career in the legal profession and lived off a handsome yearly income of roughly £50,000 from his citrus orchards which allowed him to spend most of his time in a 200 square metre house in Lahore's model town colony with his wife, three children, a driver, a cook, and a 10-year-old maid from the village. Because he had time on his hands his uncle Chowdri Nawaz Ali, a several times member of the provincial assembly, appointed him to help me settle down in the village.

During my first week in the village Mahmood Abbas Gondal took me around the area of Bek Sagrana in his chauffeur-driven car, in the company of his childhood servant Bilal Mirasi. Bilal Mirasi no longer worked permanently for Mahmood Abbas but occasionally offered his services to him when needed. Mehmood often asked Bilal to accompany him when he went to visit friends and relatives both around Bek Sagrana and beyond. During these first few days Bilal accompanied us everywhere. While Mahmood was introducing me to his relatives, Bilal generally sat in the background in the company of other servants. In the mornings Mahmood would ask him to dust his shoes and then throughout the day he would be called upon to bring glasses of water, snacks, tea or to buy cigarettes in a nearby shop. In the evenings Bilal would often provide Mahmood and his friends and relatives with a bit of entertainment by recounting some amusing local gossip or some unusual event that he had witnessed. He told stories about local cattle thieves, drug addicts, elopements, and the Koreans who had come to build the motorway connecting Islamabad to Lahore in the 1990s – who he claimed to have seen eating forbidden (*haraam*) turtle and pig meat.

The first people that Mahmood took me to visit were his maternal uncles Chowdri Nawaz Ali and his younger brother Chowdri Mazhar Ali. Chowdri Nawaz Ali had moved to a nearby village where he had more than half of his lands, and Chowdri Mazhar Ali had moved to a dera a mile to the east of the village. Chowdri Nawaz Ali Gondal was an influential member of the Pakistan Muslim League Noon (PML-N) and his younger brother,

---

[7] Both persons' personal names and castes have been changed to protect the identity of informants.

Chowdri Mazhar Ali, was the party president for the district of Sargodha. They were once the most powerful Gondals with links to Bek Sagrana, but since General Musharraf's military coup in 1999 they had been in the opposition. Chowdri Mazhar Ali had even been jailed for a year shortly after the coup, and Chowdri Nawaz Ali had had to cease his political activities and temporarily cut his ties with Nawaz Sharif's party (see chapter two). The result was that their local rivals had been able to gain power at their expense. Most notably, their cousin and fierce rival, Chowdri Abdullah Gondal,[8] had gained considerable ground against them by aligning himself with the pro-government faction. As union council nazim and the local power broker for a larger pro-government coalition, he had become the most influential patron in the village of Bek Sagrana. However, Chowdri Mazhar Ali and Chowdri Nawaz Ali were confident that the tide would turn, as it frequently did in Pakistani politics, and that they would sooner or later return to power and put Chowdri Abdullah back in his place.[9]

Like Mahmood Abbas, both of them thought it strange that a Westerner should want to research a rural backwater such as theirs. After all, they argued, all Pakistanis wanted to go to England and to America to study and do research and I was doing the opposite. When I explained that I was interested in researching feudalism for my PhD in social anthropology they proudly asserted that I had come to the right people, but they remained doubtful as to what I could possibly gain from it. Rather than reacting defensively to my mention of feudalism they proudly asserted that they were 'feudals'[10] and, waving towards the horizon, told me that all of the land in sight belonged to them. They told me that as their guest no one in their area would touch me and that even the local police would salute me. Thus despite their doubts as to the purpose of my stay in the village, they offered me their hospitality and protection.[11] Over the next few months it became

---

[8] Chowdri Abdullah was also an affine of their by virtue of being married to Chowdri Nawaz Ali and Mazhar Ali's half-sister.

[9] In 2008, two years after the end of my fieldwork, the tide did turn and they regained the upper hand over Chowdri Abdullah.

[10] They used the word in English.

[11] I soon realised that such 'protection' was necessary in order to avoid harassment at the hands of the local police and local toughs. Whenever a police officer stopped me, the fact that I was with the Gondals meant that rather than harassing me they invited me to have cups of tea. Moreover local toughs who were initially hostile to my presence in the village accepted my presence there when it became clear to them that I was a guest of the Gondals.

clear to me that both Mahmood Abbas and his uncles believed that I was a spy, but this somehow never interfered with their hospitality towards me.

After several long days lounging on charpais, drinking tea, smoking hookahs, and meeting a bewildering number of multi-related Gondals, I was eventually dropped off at the place where I would initially lodge in the area of Bek Sagrana. Mahmood Abbas offered to house me at his father's dera, which was also a Sufi lodge built in the 1980s. This was a place where members of a particular branch of the Qadiri Sufi order (*silsilah*) occasionally gathered to perform *zikr*[12] and where Mahmood Abbas's father, Sufi Ahmed Abbas, held the monthly celebrations of Gyarvi Sharif commemorating the date of Sheikh Abdul Qadir Gilani of Baghdad's union with God. His father agreed to my staying there, and Mahmood Abbas appointed Bilal Mirasi to help me settle down in the village and also to bring me my three daily meals from Sufi Ahmed Abbas's house.

I initially worried that this arrangement and my entry into the village through the Gondals might make it appear to villagers that I was on the side of the landlords. I thought that this might prevent me from winning the trust of *kammi*s (members of the various menial and artisanal castes) and poorer villagers. However, with time my worries on this matter began to subside for several reasons. To begin with, Mahmood Abbas Gondal and his elder brother lived in Lahore and only came back to the village once every two weeks. Sufi Ahmed Abbas, Mahmood Abbas's father, was a pious recluse who initially tolerated my presence at the lodge but who was suspicious of me and who therefore kept his distance. This all meant that as soon as Mahmood Abbas left the village I spent all of my time with members of the various dependant kammi households that clustered around Sufi Ahmed Abbas's lodge. Bilal Mirasi kept up a degree of formality with me for the first few days but soon started to relax when he saw that I didn't treat him the way that landlords usually treated their servants. Within a couple of days he started openly smoking marijuana in front of me, took me along to meet all of his friends, and freely spoke to me about the problems that he faced with his Gondal masters.[13] Moreover because the Sufi lodge was a place where villagers from different social backgrounds gathered to

---

[12] The word 'zikr' literally refers to the act of remembering, and in a religious context it refers to the act of remembering God, usually through the recitation of his 99 names.

[13] After about ten days he asked me whether he could take leave in order to work in the citrus harvest, and assigned one of Sufi Muhammad Hayat's child servants to bring me my meals. Since he lived at Sufi Muhammad Hayat's dera, I continued to see him frequently after he left.

worship, socialise, eat, and rest, it was a good place to meet people.[14] A few of Sufi Ahmed Abbas's bonded labourers had breakfast and dinner at the lodge, and by eating with them on a regular basis I not only gained their trust but also their friendship. Living at the Sufi lodge also gave me a chance to closely observe the everyday interactions between Gondal landlords and their servants, labourers and clients. Sufi Ahmed Abbas and his sons would sometimes sit on charpais under a large rubber tree outside the lodge and direct farm operations, resolve people's disputes and dispense patronage to clients.

The fact that the Sufi lodge was neutral territory and Sufi Ahmed Abbas didn't take sides in the factional rivalries and conflicts that raged in the village was a further advantage, since it meant that I was able to meet and talk to the leaders of both village factions while I lived there. Because I was Sufi Ahmed Abbas's guest I was able to meet Chowdri Nawaz Ali's biggest local rival, his distant cousin Chowdri Abdullah Gondal, both at the Sufi lodge and at his dera on several occasions. When I later had to shift to the dera of Chowdri Nawaz Ali's youngest brother, Chowdri Arif, I ceased to have the privilege of neutrality.[15] Chowdri Arif was allergic to Chowdri Abdullah's very name and made it clear to me that he wouldn't tolerate me socialising with him regardless of my research interests. Nevertheless, during the initial months at Sufi Ahmed Abbas's dera I was able to visit Chowdri Abdullah at his dera on several occasions and observe how he dealt with village affairs. Subsequently, even though I could no longer visit him I was able to gather a great deal of information about him and his brothers through his allies and servants as well as through villagers, for whom Chowdri Abdullah's character and activities provided an inexhaustible topic of conversation.

Chowdri Nawaz Ali's younger brother owned around 30 acres of citrus orchards, considerably less than Chowdri Mahmood's 150 acres, but was nevertheless able to live in Lahore with two wives, four children and a child servant and to drive around in a two door Suzuki – he was one of eight Gondals to own a car and could boast of having been the first to have purchased one. Since he was away most of the time I also spent most of my evenings in his house in the company of his farm servant and his friends and relatives. During the day I would either visit the labourers who lived at Sufi Ahmed Abbas's dera to chat with them or to accompany them to the fields

---

[14] The correlative disadvantage was that the place afforded me no privacy, but I eventually got used to this.
[15] Eventually Sufi Muhammad Hayat decided that I should live elsewhere. Friends in the village told me that he had become uncomfortable about lodging a non-Muslim in a place of Islamic worship.

and citrus orchards, or I would cycle to the village where I had established a close friendship with the two nephews of the village Imam shortly after my initial arrival.

Unlike elite Gondals, the village Imam and his family resided permanently in the village and they were an integral part of the village community. Although they were wealthier than most villagers – they owned 2.5 acres of land, a motorcycle, a washing machine, two bicycles and lived in a *pakka* house – the difference between them and ordinary villagers was far lesser than that between villagers and people like Chowdri Mahmood Abbas. Because of this and because of their reputation for learning, piety, honesty, and charity, my association with them helped me become accepted by many villagers who initially associated me with the landlords and who were uncomfortable about the presence of a non-Muslim and potential spy in their village.

Villagers were very concerned about whether Islam permitted them to befriend, interact and share food with a non-Muslim. Many of them also thought I might be a spy. Mullahs and school teachers often propagated a state-inspired ideology according to which Pakistan was besieged by non-Muslim enemies who didn't want to accept the existence of a nation built not only to defend South Asia's Muslims but the entire Islamic world. They taught that it was because Muslims were besieged by infidel enemies that Pakistan had acquired the atom bomb to defend them. Westerners were depicted as particularly threatening because in addition to their overt belligerence they were spreading immorality and selfishness (*khudgarzi*). Westerners were alleged to be solely concerned with the satisfaction of their individual material and bodily desires making them no better than animals (*janvaar*). Their acquisitiveness had allowed them to obtain considerable material wealth but their entire social order was built on the unsustainable basis of selfishness.

As this book will show, villagers felt that 'Western' self-indulgence and selfishness had penetrated Pakistani society and had seriously undermined the traditional moral order and social fabric. One of the first things that the village Imam's nephew told me was that soon Pakistanis would abandon *purdah* and engage in free sex like Westerners who had no shame and who despite their technological feats and material wealth were no better than animals. Because I was a Westerner I had to spend a great deal of my time in Pakistan trying to convince people that I wasn't an agent of moral degeneration.

Villagers believed that people, and landlords in particular, no longer shared their wealth because they put money and consumer goods above social relations and even God. Moreover they believed that force and fraud had become the prerequisites of power and wealth. To obtain these people stole, killed, kidnapped, bribed, sold adulterated goods, trafficked heroin, and engaged in bootlegging. The resulting social strife was seen as God's just punishment. There was a chronic insecurity of life and property coupled with a pervasive sense of mistrust in both people's honesty and in the quality of purchased consumer goods. Both people and goods were constantly referred to as second rate (*do nambar*) indicating that they were untrustworthy. Everything from milk to electronic goods and pharmaceuticals were held to be of poor quality because of tampering and adulteration. Moreover because people had lost all sense of shame (*ghairat*) by succumbing to their carnal (*nafsiati*) lusts and desires, sexual promiscuity was rife and young men drank openly in village streets, hurling abuse at people and molesting unsuspecting girls in alleyways and fields at night.

Given that 'Western' moral values and political interventions were held to be the cause of the current moral disorder it is unsurprising that people were weary of hosting a westerner. Only a couple of days after arriving in the village, a young school boy threw stones at me and called me a *kaffir* as I cycled into Bek Sagrana.[16] A couple of days later an octogenarian lady apparently walked several miles for the sole purpose of warning my hosts that I was a spy paving the way for an 'English' (*angrez*) invasion of Pakistan. Acquaintances often told me that people in the area were saying that the Gondals should throw me out of the village. With time I also learnt that many people had questioned Sufi Ahmed Abbas about his decision to allow a ritually impure (*napak/pleet*) infidel to reside in a place of worship. Although his version of Sufism made him relatively tolerant of religious diversity he wasn't happy about the fact that his Islamic credentials were being questioned due to my presence in his lodge. This eventually led him to find a pretext to kick me out and forced me to find another place to live in.

So long as I spent my days with prominent Gondals I was largely insulated from the controversy that my presence was generating. This is because few people had the temerity to raise doubts about powerful landlords and their guests. Landlords could openly pursue amorous affairs, drink, take drugs, and even follow highly unorthodox versions of Islam without being questioned by ordinary villagers, so no one was going to openly raise questions if they wanted to host an infidel. It was only once Chowdri Mehmood Abbas was no longer around that the issue of whether people could eat with

---

[16] When a school teacher reprimanded the young boy and told the boy that I was a guest the boy apologised and said that he had thought I was an 'angrez' and not a guest; the school teacher didn't bother to explain to the child that I was both an 'angrez' and a 'guest'.

me, or even socialise with me, started to arise. Fortunately I befriended the nephew of the village Imam – the only English speaker in the village other than Chowdri Mahmood Abbas and his brother – within days of my arrival. His uncle told me that as a person of the book (*ahl-e-kitaab*) people could eat and drink with me if they wanted and then declared this to assembled villagers at the mosque on the first Friday after my arrival.[17] This event, and my subsequent close association with the Imam's family made my presence in the village acceptable to the majority of villagers.

Their interest in and intimate knowledge of village affairs also meant that they were invaluable informants who took an active interest in my research. Their scrupulous commitment to telling the truth (*haqiqat*) and to speaking out against immorality, fraud and injustice meant that they were often willing to provide me with information about which others were more secretive. Thus I quickly came to learn details about the local drug trade, bootlegging, buffalo theft, electricity theft, and the various forms of embezzlement of government funds and appropriation of public office for personal use that took place in the village. Moreover whenever I needed to confirm certain facts they were often willing to introduce me to concerned parties who might be ready to impart information to me. They also helped me carry out minor surveys which would have been impossible for me to do without their support because of people's fear that I might gather information that could one day incriminate them with the authorities – as this book will show, people were extremely weary of state authorities which they experienced as exploitative and sometimes violent. Crucially, they helped me learn Punjabi and – initially – translated bits of conversations in the local Punjabi dialect into Urdu for me. After more than six months I was able to understand a considerable amount of Punjabi but continued to conduct fieldwork in Urdu – which everyone understood well, and which almost all young men spoke.

When my Gondal hosts, Chowdri Mahmood Abbas and later Chowdri Nawaz Ali's younger brother, came to the village (roughly every fortnight) I abandoned my daily routine to spend time with them. This was not only necessary because, as my hosts, they expected me to spend time with them when they came to the village, but also because it served my research

---

[17] The Imam asked me to attend the Juma prayers that Friday and told me to sit at the back so as not to interfere with people's prayers. After his sermon following the prayers he introduced me to the assembly, told them that it was acceptable for them to befriend me and finally asked them to pray for my conversion to Islam. He told them that they should be exemplars of piety and morality in order to inspire me to convert.

interests. By spending time with them I was able to gain insights into the political intrigues, rivalries and enmities which often dominated their conversations. My good relations with them meant that they introduced me to their friends and political allies throughout the district. They also introduced me to lawyers, judges and police officers who were sometimes willing to discuss local politics with me. On a few occasions they also invited me to Lahore, where I was able to see how they and the servants that they took with them from the village lived while there. There were times when I also visited Chowdri Nawaz Ali and Chowdri Mazhar Ali to carry out informal interviews with them. Their sons were particularly eager to show me around and to discuss local, regional and international politics with me.

The people I had almost no access to were the women. The Gondals maintained strict purdah with outsiders and only close relatives and dependants from the village were allowed to see the women of their households. When I inquired about marriage alliances, they were willing to tell me about the men that their female relatives had married, but they never referred to the women by name and never spoke about them. The little information that I was able to gather about them came from servants who told me about which *chowdranis*[18] they liked and disliked. Nor did I even have much access to lower caste women who couldn't afford to maintain such strict purdah. Although I often saw them when I visited kammi friends at their houses, I almost exclusively spoke to the men. Given that the presence of a non-muslim and potential spy in the village already caused controversy, I didn't want to make matters worse by even raising the possibility that I was a threat to the modesty (*hayaa*) of village girls. I quickly learnt that even an innocent greeting to a female labourer I saw on a daily basis walking down the street was capable of generating gossip so I was careful to only speak to women if they were accompanied by their male relatives.[19]

Finally, although Bek Sagrana was my principal field-site, important aspects of my research were informed by meetings with politicians, judges, bureaucrats, politicians, journalists, and NGO workers in both Lahore and Islamabad. My research was also informed by visits where I met people in other villages throughout the district and beyond. Although these people

---

[18] A term for chowdri women.

[19] On one occasion, near the beginning of fieldwork, I briefly paused to talk to some women who greeted me on my way to the village and the next day there were already rumours about my trying to seduce village women. My friends promptly advised me not to speak to women unless their husbands happened to be around. The fact that I was a Westerner appears to have made this even more of an issue for me since people initially assumed that I would be sexually promiscuous.

do not form part of the material presented in this book, my conversations with them helped me to formulate lines of inquiry and gain insights into the broader relevance of the material that I faced in the village. As Clifford Geertz argued 'The locus of study is not the object of study. Anthropologists don't study villages . . . they study *in* villages' (Geertz 1993 [1973]). The village therefore provided me with a place to explore broader political issues that are relevant to academics and policymakers working on Pakistan. Moreover, although it is true that villages in South Asia are no longer political, social and economic units they do remain important contexts for understanding social and political change since, as Jeffrey, Jeffrey and Jeffrey (2008) argue, 'employment opportunities for most rural people remain limited to the village area, most villagers depend upon services located in the village, and the village remains a powerful imaginative social unit' (ibid.: 24). I would add that people also remain attached to villages because of the fact that relatively spacious and cheap accommodation is still affordable there.

Starting in July 2001, I spent a total of about 25 months in Pakistan over the course of several trips, and I spent 20 of those months in Bek Sagrana. I initially stayed in Bek Sagrana for six months between December 2004 and May 2005. Due to visa problems I was forced to return to the United Kingdom and was only able to return to Bek Sagrana in Febuary 2005, after which I spent a further 14 months there until the beginning of May 2006. During that time I spent periods of up to a week in either Islamabad or Lahore every two months. I spent my first three months in Pakistan (between July and September 2001) working in an NGO in the Hazara district of the North West Frontier Province. I then spent two months between December 2002 and January 2003 travelling around the Punjab looking for prospective field sites.

Overview of Chapters

Chapter one begins by locating Bek Sagrana within the broader political, economic and historical context of the Punjab and, more specifically, of the Sargodha district. It then provides a short history of the village and illustrates how the Gondals have consistently asserted themselves through the use of force. The chapter then provides a brief sociological overview of the different status groups that make up the village and of how their means of livelihood has changed over the years. It illustrates how the village has ceased to be the principal centre of economic and political activity as both landlords and their landless dependants increasingly earn their livelihoods beyond the village. While the latter increasingly earn their livelihoods as casual wage labourers, the former have been able to buttress their local political and economic clout by getting involved in politics as well as by obtaining government and private sector jobs and, in a few recent cases, by migrating abroad.

Chapter two explores how despite agrarian transformation, landlords continue to use debt bondage to exploit labourers for both their political support and labour. I argue that the decline of the traditional division of labour – characterised by interdependence between landlords and the various occupational groups known as kammis – and the modernisation of agriculture has Turned landless villagers into a proletariat dependent on intermittent casual labour for its subsistence. While attachment to landlord households was once the default position for most, the contemporary default position was to be a free wage labourer, although during times of crisis many were forced to attach themselves to landlord households in order to get loans. Because of low and irregular wages, labourers were unable to make savings to cover large wedding expenses or emergency medical ones and had to turn to landlords to obtain them. I argue that through these debt-relations the Gondals were able to bind an increasingly independent workforce. Moreover I argue that the effect of using the kinship ties of attached labourers and servants to make them service their debts was that it displaced potential class conflict onto the kinship group of the labourers. The overall result was that labourers remained divided and therefore unable to collectively bargain for better pay and work conditions.

Chapter three begins with a historical overview – from the colonial era to the present – explaining how landed power became entrenched in the Punjab. Having set the broader context of landed power in the Punjab, the chapter zooms into the village in order to examine the strategies and mechanisms used by Gondal landlords to reproduce their power. It asks whether electoral politics mainly benefitted the landed elite or whether political competition led to an increase in clientelistic provision to the poor. I argue that the main beneficiaries of patronage tended to be the Gondals themselves and that only a minority of the poor substantially benefitted from it. Moreover, I argue that even if we accept that some of the poor did benefit, clientelistic provision tended to undermine the state's role as a provider of public goods – such as health care, justice, housing, and even education – because it occurred along the particularistic lines of kinship, clan and faction. By effectively privatising access to and control over the state along these lines politicians also reproduced traditional concentrations of political and economic power and even created new ones. Thus contrary to Partha Chatterjee who believes that popular freedoms advance through the ad hoc clientelistic practices of what he calls 'political society', I argue that these practices disempower the common man and consolidate existing forms of domination.

In the chapter's conclusion I suggest that Pakistan's failure to establish impersonal governance and programmatic politics is the result of a combination of state level authoritarianism and of the presence of powerful landed interests on the ground. I argue against the notion that the state is unable to implement its policies because, in Migdal's (1988) terms, Pakistan has a strong society and a weak state – a claim made in different ways by both Anatol Lieven (2011) and Matthew Nelson (2011) who argue that kinship ties undermine both policies and the legal system.

Chapter four examines the role of the extended lineage/clan (*biraderi*) as a unit of political mobilisation in the Punjab. Drawing upon Barth's work on agnatic rivalry among the Swat Pathans, I argue that competition over the control of land and – more importantly – over the control of votes have the effect of obliterating the unity of extended lineages. Thus I argue – contrary to both popular and academic literature – that extended lineages don't determine people's political alliances. Far from being fixed by primordial ties, people's political alliances in Bek Sagrana were highly instrumental and fluid. Focussing on village leaders, the chapter shows how people joined forces along factional lines that cut across ties of caste and class, and how they did this on the basis of shared enmities.

Chapter five returns to how landlords in Pakistan undermined state policies. It does so by examining how they appropriated General Musharraf's devolved government scheme; a scheme that ostensibly aimed to transfer power from the elites to the vast majority. The programme – designed in consultation with a variety of international aid agencies – was premised on the idea that the Pakistani state's lack of accountability was due to a tradition of bureaucratic authoritarianism dating back to colonial times. To improve accountability it was therefore necessary to get rid of this tradition by creating new, local government structures. The problem is that the scheme failed to acknowledge the facts that military rule and landed power were also responsible for the state's lack of accountability.

The chapter begins with a history of devolution programmes in Pakistan and looks at how military regimes used them to buttress their power by creating a loyal class of politicians as well as by substituting democratisation at the provincial and national levels with democratisation at the local level. It then goes on to examine the Union Council Nazim elections of 2005, and looks at how the Musharraf regime systematically rigged them.

The final chapter starts with a discussion of how Islamisation in Pakistan served to combat the left and to divert attention away from issues of social justice and of how it ultimately backfired by giving birth to the militant movements who are now challenging the writ of the state. The Islamisation of Pakistan – intensified under General Zia ul-Haq's rule – was part of the cold war struggle against communism and served to eradicate the left from the country's political scene. Because of Islamisation, issues of social equity and justice disappeared from the political agenda of most parties in the country. This was reflected locally in the fact that political debate tended to focus on the individual morality and piety of leaders rather than on issues of social equity. Locally Islamisation meant that debates on Islamic morality were always in the foreground.

I then move on to examine the local flavour of Islamisation in the context of Bek Sagrana. The Islamic tradition predominant in the rural Punjab was a Sufi one that centred on saints who mediated between the divine and the secular realms. As David Gilmartin (1984) has shown, Sufi Islam in the Punjab was closely tied to the local landed power structure. In this chapter I examine the extent to which Sufi-inspired ideas served to legitimise social inequality. Deobandis and Islamic modernists in Pakistan both decry living saints who double-up as powerful landlords as obscurantists and blame them for the country's social and political backwardness. Allama Iqbal once termed the form of religiosity around living saints 'Persian mysticism',[20] and blamed it for the allegedly blind submission of the masses to an elite claiming spiritual pre-eminence. He believed that this was the root cause of feudalism in Pakistan. More recently, Abdellah Hammoudi (1997) has made a similar claim that authoritarianism in Morocco and the Middle East is rooted in a deeply ingrained respect for authority derived from the authoritarian relationship between Sufi masters and their disciples. He argues that large-scale organisations in Morocco such as government bureaucracies and political parties are governed by criteria of personal allegiance and faithfulness that replicate those governing the interactions between masters and their disciples in Sufi brotherhoods. In this chapter I examine whether his and Iqbal's claims are valid in the context of Pakistan, and if not whether it is nevertheless still fair to say that Islamisation in Pakistan has predominantly served elite interests.

---

[20] Persian mysticism created a 'Sufi' spiritual aristocracy in the form of pirs pretending to claim power and knowledge not accessible to the average Muslim.

1

THE VILLAGE

District Sargodha

The Punjab is home to the five rivers that give the province its name.[1] The river Jhelum serves as the Punjab's Western boundary with the Sutlej serving as the region's Eastern boundary. In between these two rivers lie the Chenab, the Ravi and the dry bed of the Beas River.[2] The areas between these rivers, known as *doabs*,[3] constitute some of the most fertile land in all of Pakistan. The research for this book was carried out in the village of Bek Sagrana in the central Punjabi district of Sargodha located in the Jech doab, demarcated by the Chenab River to the East and by the Jhelum River to the West. According to Wilder, the Central Punjab is the Punjab province's

> ... political, economic and cultural centre. It is the most urbanized, industrialized, agriculturally productive, and densely populated of the four regions of the Punjab. The sixty-one National Assembly Seats of central Punjab comprise more than half of the total seats of the Punjab and more than a quarter of the entire country's seats. The key to success in Punjab politics and to a considerable extent Pakistani politics lies here.
>
> (Wilder 1999: 34–37)

---

[1] Punjab derives from *panj ab* meaning five waters.
[2] The water of the river Beas along with some of the water from the Sutlej and Ravi rivers was diverted to India as part of the Indus Water Treaty between India and Pakistan in 1960.
[3] The term 'doab' literally means 'two waters'.

Sargodha is one of the districts that comprise the central Punjab, with five national assembly seats and eleven provincial assembly seats.[4] Subsequent to canal colonisation that began in 1885, the area became one of the most agriculturally productive areas in Asia[5] as well as the most densely populated area of the Punjab.[6] Fertile land and an extensive irrigation system contributed to making the central Punjab the centre of Pakistan's green revolution. In addition to being the richest region of the Punjab agriculturally, it is also the most industrialised one. In 1989, 71.9 per cent of the Punjab's industry was located in the central Punjab, principally concentrated in the cities of Lahore, Faisalabad and Gujranwala (ibid.: 40).

As a result of canal colonisation the Sargodha district became a major producer of cotton, wheat, barley, maize, millet, rapeseed, and pulses. By 1947 the Sargodha district had become one of the largest agricultural hubs in Asia, with a major market for grain (particularly wheat) and 10 large and well-equipped cotton ginning factories. At the time of independence, when Pakistan emerged from partition with a poor industrial base, the district was a major contributor to Pakistan's tax revenue and a major focal point of foreign exchange. Agriculture in the district received a further boost during the green revolution starting in the 1960s. During the 1980s the extensive irrigation system of the district made it possible for many landlords to start substituting citrus cultivation, requiring significant amounts of irrigation, for cotton cultivation. By 2004 the Sarghoda *tehsil* of Bhalwal, where the research for this book was carried out, came to have the highest density of citrus orchards in the district and was often referred to as the 'California of Pakistan'. The production of citrus was not only more profitable than cotton, and boosted the district as a centre of foreign exchange, but was also significantly less labour intensive. As chapters three and four will illustrate, this allowed wealthy landlords to spend less time supervising agricultural activities and more time in cities such as Sargodha and Lahore, where their children could obtain a better education than in their home villages. This significantly affected the quality of patronage ties between landlords and villagers.

---

[4] These districts include Gujranwala, Gujrat, Sialkot, Narowal, Lahore, Kasur, Sheikhupura, Sargodha, Faisalabad, Toba Tek Singh, Sahiwal, and Okara.
[5] See Ali (1988) for a comprehensive account of canal colonisation in the Punjab.
[6] In 1993 the population of the central Punjab was 34,500,000, more than half the Punjab's population (Wilder 1999: 38).

Given the highly unequal distribution of land and access to formal state institutions (which were to a large extent the legacy of the colonial practice of indirect rule through landed notables), the benefits from both the green revolution and the introduction of citrus orchards accrued principally to the landed elites. According to World Bank (2002) estimates, less than half of rural households in Pakistan own any land, and that more than half of rural farm holdings are of less than five acres (ADB 2006: 46) and account for 16 per cent of Pakistan's total farm area. On the other hand, Malik (2005) reports that in 2000 only 5 per cent of farms were 25 acres or more in size but that they accounted for 38 per cent of all owned land. Moreover statistics for the Punjab from 1976 show that landowners with more than 50 acres of land accounted for more than 18.2 per cent of all land owned. Hussain (1989) and Zaidi (1999) point out that during the green revolution it was principally the large farmers who could get credit to finance the use of new inputs and technologies. These farmers not only possessed substantial collateral in the form of land but also had privileged access to the state distribution of inputs and technologies through their connections with politicians and bureaucrats. On the other hand, tenants and smallholders had to obtain credit through informal institutions, often through landlord farmers, who required them to repay in kind and who often valued their produce at rates that were below market rates. In addition, access to the market in remote areas was often controlled by landlords who owned trucks and marketing outlets, and who could thereby extract a surplus from local smallholders and tenants. The net result was that smallholders derived little benefit from the green revolution and that many even ended up highly indebted and were forced to sell their land. This trend continued well after the initial onset of the green revolution. Hussain (2003) claims that many such smallholders were increasingly being forced to sell their land, and that between 1990 and 2000 as many as 76.5 per cent of the extremely poor and 38.9 per cent of the poor had done so.

Sharecroppers were also negatively affected by the green revolution. Ahmad (1977) reported that prior to the green revolution and to the 'tractorisation' that came with it, landlords in the Sargodha village of Sahiwal begged tenants to cultivate as much land as they could in exchange for 50 per cent of the harvest. The reason for this was that many landlords were unable to organise the cultivation of their lands themselves due to the vast number of bullock teams and workers that this would have required. Chakravarti's (2001) work on Bihar shows that whereas a single bullock team was able to prepare 0.42 acres of land in six hours, a tractor could prepare 25 acres in 20 hours. According to one estimate in

Pakistan each tractor introduced in the 1960s displaced between 9 and 12 labourers (Sayeed 1996: 276).

Thus, 'tractorisation' meant that landlords could replace a large number of tenants and their bullock teams by a single tractor driver. 'Tractorisation' also meant that a lot of land dedicated to the cultivation of fodder for the bullock teams could be turned over to other crops, which dramatically speeded up the turnover of crops in single fields. Lastly, it allowed wealthy landowners to do away with tenants who, under both General Ayub Khan's and Zulfikar Ali Bhutto's land reform programmes, might have decided to claim legal title to the land they cultivated. The result was that between 1960 and 1990 the total area of land in the Punjab cultivated by sharecroppers declined from 37.2 per cent to 14.2 per cent (Zaidi 1999: 42). Finally, in districts like Sargodha, the introduction of citrus orchards in the 1980s also played a significant role in displacing agricultural tenants.

In this way many former smallholders and tenants, as well as village artisans whose goods were to a large extent replaced by mass produced ones, joined the ranks of the mass of 'unorganised and unprotected workers' (Breman 1996: 2) bogged down in the transition from agrarian to industrial production. Workers pushed out of agriculture and cottage industries by mechanisation were not formally incorporated into a growing industrial sector. Instead the majority of this work force was employed in a variety of petty trades, services and casual labour in the agricultural sector, the industrial sector and in construction work. Construction work was a particularly important source of employment. The 1998 Government Census claimed that 35.8 per cent of the district's population was employed in construction, followed by 31 per cent who were employed in agriculture. What the census statistics do not show, as will be described in chapter two, is that the majority of labourers were casual, and moved between agriculture, construction and industry principally located in neighbouring districts.

The Gondals of Bek Sagrana

Prior to canal colonisation, most permanent settlements on the Jech doab were around the banks of the Chenab and Jhelum rivers. Here irrigation was principally carried out through the use of wells (*khus*) and occasionally through inundation canals. Away from the rivers, scant rainfall made the Jech doab largely unsuitable for settled agriculture despite its rich alluvial soils. These inland areas were sparsely populated by semi-pastoral people (*charaghs*) who kept livestock and practiced limited single cropping agriculture. Near the banks of the Chenab River in the eastern part of the Jech doab, where the research for this book was carried out, the two dominant semi-pastoralist Jat clans were the Gondals and the Ranjhas.[7] Once canal colonisation was underway the colonial administrator Malcolm Darling reported that these pastoralists, or *janglis*[8] as he pejoratively referred to them, were poor cultivators 'like all primitive folk who have an abundance of land' (Darling 1934: 14). He claimed that 'if the jangli is not a good farmer, he is at least a good sportsman. Faction and feud are rife in his villages, and he likes to settle his quarrels in old fashioned ways without recourse to court and police' (ibid.: 15). Moreover he claimed that 'once there was hardly a *zaildar* who was not in the cattle-thieving business, and even now it would be difficult to find anyone of any prominence who had not a relative or two connected with it' (ibid.). Canal colonisation made perennial agriculture possible and transformed these sparsely populated doabs into the agricultural heartland of the Punjab. Settlers (abadkars) were brought in from other areas of the Punjab and were given titles to canal-irrigated land. Most of these people settled in newly built, planned villages known as *chaks*. The geometry of these villages was one of squares and straight lines, and reflected the colonial government's self-professed civilising mission which aimed to create modern and enterprising farmers (Gilmartin 2004). Groups that were native to the region such as the Gondals and the Ranjhas were also given titles of landownership over the newly irrigated land and became settled agriculturalists. However, many of them, including the Gondals, continued to live in the old, unplanned village settlements.

The village of Bek Sagrana is situated to the east of the district towards the banks of the river Chenab. For the village landlords the trip by car to the city of Sargodha took around forty minutes, while the trip to Lahore, along the Korean-built motorway completed in 1997, took about two hours.

---

[7] Both the Ranjha and Gondal populations were spread across a large geographical area that spanned from the district of Gujrat to the east and across the Jhelum River to the west. Within both of these Jat clans there sub-clans that often acquired their names from a patrilineal ancestor. Thus, for example, the descendants of a man called Kala Gondal were known as the Kaliana Gondals. Alongside the Gondals and *Ranjhas* other dominant landowning clans in the area of Sargodha included *Pathans*, *Rihans* and *Nissuwanas*. To the West of them, in the Jhelum valley, the dominant landowning clans in the district included *Bhattis*, *Khokars*, *Mekans* and *Jhammats* as well as *Baloches* and *Sayyids*. *Tiwanas*, *Noons* and *Awans* were other important landowning clans to the East of the District.

[8] The term literally refers to 'people of the jungle' and pejoratively indicates savagery and lack of education. In Urdu the term *jangal* is used not to refer to tropical rainforests as in English but to any wild and uninhabited area of land.

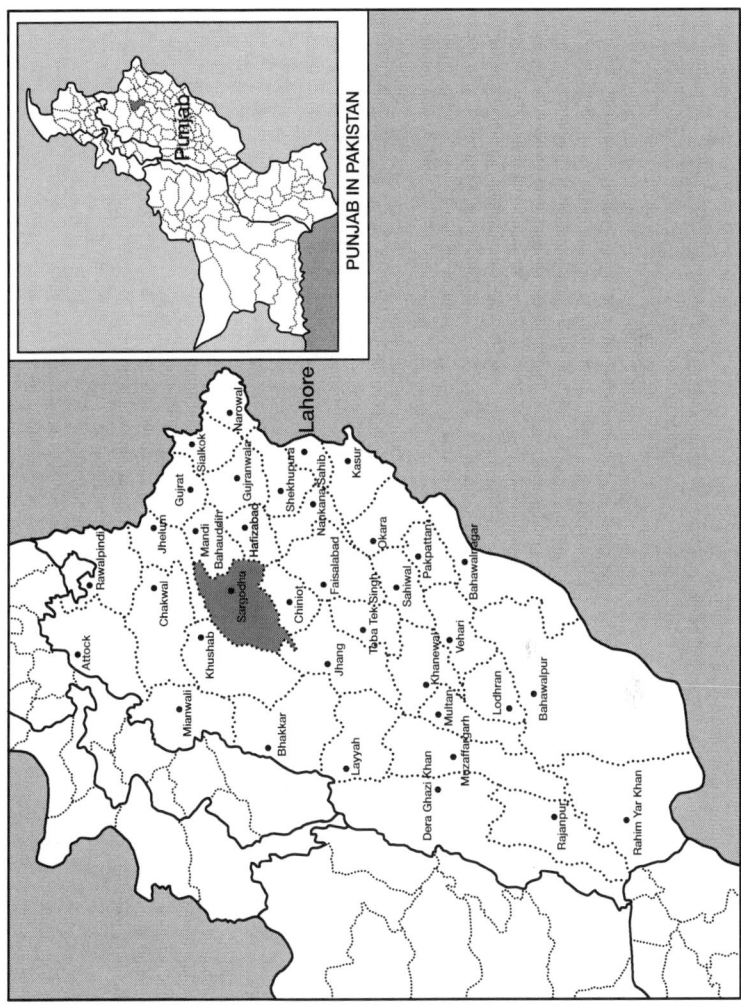

Map 1.1 Map of Pakistan

Source: Courtesy of http://geology.com/world/pakistan-satellite-image.shtml

For poorer villagers without cars or motorcycles, the trip to Sargodha took an hour and a half on a cramped bus with a deafening musical horn.[9]

The history of the village was one in which the Gondals repeatedly asserted their dominance against other clans and even against the state. Villagers claimed that Bek Sagrana had once been on the banks of the river situated to the east of the village. With time the shifting course of the river had moved further west, and what was now left of its former course formed a marshland (*buddhi*). Some villagers claimed that the river had moved west as a result of the prayers of a local village pir whose shrine lay to the east of the village cemetery facing the lowlands. When the river moved, the Gondals allegedly moved with it because of the need for irrigation water. In their absence the *Sagranas*, a local cultivator lineage (biraderi), had overtaken the village. Subsequently one of the ancestors of all of the present day Gondals in Bek Sagrana, around whom a great deal of legend revolved, decided to resettle in the village and re-conquered it by force and cunning. His descendants boasted about how this ancestor, known as *Kala Gondal*, or Black Gondal, had called for a gathering with leading Sagranas and had got his men to ambush and kill them as they were making their way to the meeting.[10]

Much later the great-grandfather of all of the Gondals in Bek Sagrana had murdered a colonial revenue officer (*tehsildar*) who was alleged to have barged into people's houses without respect for purdah and to have extorted money from villagers. The killer and some of his accomplices were subsequently tried and hanged. In these stories the Gondals revealed themselves as ruthless, brave and cunning people; qualities deemed essential for effective politicians. They claimed to possess these because of their fundamentally passionate (*jezbati*) nature which meant that they might act recklessly, and even cruelly, but always out of concern for the interests of kin, friends, allies, and servants.

When I began fieldwork in 2005, many landless villagers still used the lowlands near the former banks of the Chenab river as pastureland for their livestock, but in 2007 Chowdri Abdullah Gondal turned it over to intensive agriculture after forcefully capturing the land from the Makhdooms.[11] Long ago, before the arrival of canal irrigation in the second half of the 19th century, one of the Gondals had donated this land to a

---

[9] The bus was owned by the younger brother of one of the village leaders.

[10] Despite the fact that the Gondals had retaken the village and claimed that it was rightfully theirs, the village retained the name of the Sagranas.

[11] In South Asia the term *Makhdoom* is used to refer to families that descend from a saint (*pir*).

Makhdoom saint whose prayers had supposedly made God grant him a son. The formerly politically influential Makhdooms had mainly used the area for hunting wildfowl, but had lost it to Bhutto's land reforms and to capture by members of the Gondal clan. Elderly villagers recalled how the Makhdooms had brought Europeans along during their hunting excursions. However, like other erstwhile powerful aristocratic families, the fortunes of the Makhdooms had steadily declined since the 1970s and they were no longer seen hunting. Many powerful landlords had been able to circumvent land reforms by colluding with the local bureaucracy and by evicting tenants who might claim occupancy rights (see Nelson 2011: 146–54), but the Makhdoom's failed to do so because of their lack of involvement in local politics. Their unwillingness to sully themselves with politics and their small numbers in the area also meant that the far less wealthy but forceful and numerous Gondals of Bek Sagrana were able to overtake them politically and to encroach upon their land. The Gondals justified capturing land they had once donated to the Makhdooms by claiming that the person who now claimed to descend from the Makhdoom saint who had successfully interceded for their ancestor was an interloper and was in fact the descendant of a mere carpenter.

The changing nature of dominance in a Punjabi village

In the Pakistani Punjab, as elsewhere in South Asia, the state and the expansion of capitalism reduced the role of villages as centres of social, economic and political activity.[12] The political and economic clout of landlords ceased to solely depend upon their local control over the land and the labour force and came to increasingly depend upon their involvement in supra-local politics in urban centres. Moreover the livelihoods of the landless and of village artisans ceased to revolve around the village as they were compelled to seek gainful employment further afield in towns throughout the Punjab and beyond. The result was the dissolution of hierarchical ties of interdependence between landlords and villagers and their replacement with more impermanent, contractual ties. As this book will illustrate however, this by no means heralded the end of Gondal political and economic dominance.

As indicated above, the overall result of mechanisation was that a number of sharecroppers joined the ranks of the mass of 'unorganised and unprotected workers' (Breman 1996: 2) bogged down in the transition from agrarian to industrial production. In the area of Bek Sagrana the Lurkas and the Sagranas were the main landowning (*zamindar*) clans involved in

---

[12] See for example Barth (1981) and Breman (1974).

sharecropping for the Gondals and constituted 15 per cent (19 out of 123) of the village population.[13] The average landholding among the 14 Sagrana households was 1.3 acres and the wealthiest households owned four acres of land. Of the four Lurka households in the village none owned land except for one (which owned two acres). It was because of the relatively small size of their holdings that they traditionally supplemented their incomes through sharecropping. The introduction of tractors, mechanical threshers and citrus orchards combined with the threat of land reform aiming to give land to the tiller in the seventies led many Gondals to evict them and resume the cultivation of their lands.[14] Unlike kammis the fact that they possessed capital allowed some of them to open small shops, medical dispensaries and to run rickshaws among other things. Their relative affluence also meant that they possessed the educational capital necessary to obtain low ranking formal sector jobs. Some also became drivers or farm overseers for the Gondals. However the majority supplemented their incomes through wage labour; they worked around the village during the citrus harvest between December and March and then again in April for the wheat harvest but during the rest of the year work in the village was intermittent and they sought temporary work in construction, as security guards on building sites and sometimes in factories.

Similarly the livelihoods of members of menial and artisan castes (kammis) also ceased to revolve solely around the village and the Gondal landlords. The kammis included carpenters (*Tarkhan*), potters (*Kumhar*), blacksmiths (*Lohar*), cobblers (*Mochi*), bakers (*Machi*), weavers (*Julaha*), barbers (*Nai/Hajaam*), bards (*Mirasi*), drummers (*Pirhain*), and sweepers (*Mussalli*) among others.[15] They constituted 35 per cent of the village population but this proportion was much higher in clusters of houses surrounding Gondal farmsteads – known as *deras* – where all houses other than those of the landlords were kammi households.[16] Kammis traditionally rendered their services to groups of Gondal landlords in exchange for a fixed percentage of the wheat harvest and for access to fodder and firewood from their lands. These arrangements were known as *seypi* and resembled the

---

[13] Other zamindar clans who worked as sharecroppers were the *Gurus* (two households), the *Gogs* (two households), the *Khokars* (one household), the *Theims* (one household), the *Jaras* (one household), and the *Dhulayas* (one household).

[14] Under Bhutto's land reform programme land was to be given to the tillers (see Herring 1983)

[15] See Ibbeston (1993[1916]) for detailed accounts of the traditional castes and occupational groups in the Punjab.

[16] On the dera where I lived, 16 out of 18 households were kammi households.

traditional jajmani relations described in Wiser's classical text (Wiser 1936).[17] Barbers, for example, visited their patrons' households once or twice a week to shave the men and to cut their hair. They were also in charge of delivering invitations to weddings and funerals and of performing circumcisions. The principal role of Mussallis had been to visit their patrons' households when the cattle had been taken out to the well in order to sweep the cattle shed, the courtyard and the street in front of the house but also had other duties which included preparing the mud to place on the roofs of traditional dwellings, covering the chaff and straw collected during the wheat harvest with mud, making the rope that others wove onto string cots (*charpais*) and preparing cooking fires during weddings and funerals. The principal job of Tarkhans had been to make string cots, fit door frames and make certain agricultural implements but also dug the graves at funerals. Similarly all other kammis had a principal occupation and a variety of other roles to play in the traditional village economy. Kammis also occasionally had to provide the Gondals with free labour (*begar*). Finally, in addition to performing these tasks for the Gondals, kammis exchanged their services with each other. Thus, for example, a Mussalli might help a Lohar to mend his roof in exchange for repairs to his sickles (used for harvesting wheat and fodder).

In the past all kammis who lived in the village worked for the Gondals and for each other. This is partly because work that required commuting was virtually nonexistent since the nearest town, Sargodha, was 30 km away and buses weren't readily available until well into the 1970s. But it is also because under customary law – codified under British rule – kammis were bound to provide their services to the village proprietary body.[18] If kammis didn't want to do so and had no debts towards the Gondals they could be made to leave the village; in other words their duties towards the Gondals were based on residence rather than personal liability.

Arif Hassan has somewhat optimistically argued that economic change 'freed the kammis from servitude and because the kammis had marketable skills they improved their social and economic standing' (2009[2002]: xvi). Although this may have been true of kammis living near big cities where there was a market for their skills it wasn't true for those in more remote villages such as Bek Sagrana. Muhammad Azam Chaudhary (1999) shows

---

[17] See Kessinger (1974) for a description of these arrangements in an East Punjabi village.

[18] Rose writes that 'The village *abadi* (population) belongs to the proprietary body of the village and the custom assumes that nonproprietors have settled under grants from that body' (Rose 1911: 217).

how the kammis of Misalpur, near the textile producing city of Faisalabad, left their traditional occupations to work in textile factories, to set up their own power-looms and even got educated and started working in offices. On the other hand in Bek Sagrana only some kammis had successfully marketed their traditional skills and most now worked as wage labourers while some supplemented their incomes with seypi arrangements – which, however, no longer provided them with the bulk of their incomes. Moreover even in cases where kammis had prospered they hadn't entirely shed their servile ties towards the Gondals because they still depended on them for their place of residence; this was the case despite Bhutto's homestead reforms which granted people legal ownership over their houses.[19] Because the Gondals could still evict them, regardless of the law, kammis were compelled to provide them with free labour services and to work under seypi arrangements – which kammis often felt were unfavourable but which at least granted them a limited degree of economic security. Moreover, as will be shown below, residence ties also allowed the Gondals to compel them to vote according to their dictates.[20]

In Bek Sagrana the successful kammis included a Nai household, a Tarkhan household and a Lohar household. The Nai household had a member who migrated to Egypt where he worked in a barber shop and who occasionally sent remittances to help family members pay for wedding and medical expenses. One large Lohar household who had set up a flour mill and became involved in the business of installing handpumps (*nalka*s) and another large Tarkhan household who made finely crafted furniture which they sold throughout the area of Bek Sagrana. Their lucrative furniture business allowed them to buy three buffalos, a motorcycle, a television and mobile phones.[21] Despite their prosperity, members of all three households continued to provide some Gondals with free labour services and to work

---

[19] See Rouse (1983) and Gazdar and Bux Mallah (2011) for accounts of the effectiveness of homestead reforms.

[20] In the Indian context democracy and economic change have promoted the transformation of caste as a system of interdependent hierarchical social relations, as embodied by the *Jajmani* system, towards caste as a system of 'horizontal', disconnected groups, with their own distinct cultures. Dumont has described this process as the 'substantialisation of caste' (Dumont 1980: 226–27) while others have described this process as the 'ethnicisation of caste' (Barnett 1975: 158–59; see also Fuller 1996: 22–25 and Michelutti 2008).

[21] The success of the Lohar and Tarkhan households was largely due to the fact that they had large male sibling sets of seven and six respectively. This provided them with the necessary labour force to turn their crafts into successful businesses.

for them under seypi arrangements. The Tarkhans, for example, often had to fix door frames and do electrical work for the Gondal landlord whose land they had built their house on. They also still had to perform certain traditional obligations under seypi such as bringing firewood to funerals and weddings as well as grave-digging. In 2006 they planned to amass enough money to be able to resettle in the nearby market (*mandi*) town and therefore shed their residence-based ties of servitude.

The other two Lohar households and three Tarkhan households in the village hadn't established particularly successful businesses. Members of the other two Lohar households dedicated themselves to wage labour and construction work while the other three Tarkhan households supplemented their small furniture businesses with wage labour and some seypi. Members of the other seven Nai households combined barbering on seypi arrangements with wage labour and a few of them obtained jobs in barber shops in nearby market towns. All of the kammis believed that if they were paid in cash for individual jobs performed they would be better off than on seypi arrangements in which the Gondals tended to extract more labour than what they paid for at the end of the year in wheat. However in the case of the Nais, for example, competition for barbering jobs due to the large number of barbers around meant that they accepted seypi arrangements for the sake of security. The problem, as we will shortly see, is that this source of security was becoming scarce because the Gondals spent increasing amounts of time in town and many of them no longer needed their kammis' services.

Mussallis – which formed the largest group of kammis with 15 households in the village alone – Mirasis, Machis and Pirhains who didn't have particularly marketable skills, derived most of their incomes from wage labour and some of them supplemented it with seypi arrangements.[22] These kammis formed the overwhelming bulk of labourers during the citrus and wheat harvests. Moreover they were the most likely to work as domestic and farm servants for the Gondals and to become indebted to them – although other kammis and former sharecroppers sometimes did so too. The main causes of indebtedness were wedding and medical expenses which they were unable to cover due to the fact that poorly paid and intermittent wage labour made it difficult for them to accumulate savings. Work opportunities for them

---

[22] Their seypi arrangements no longer included the duty to sweep people's houses; when they did such work it was in their capacity as privately employed servants for the Gondals. Villagers were aware of the fact that Mussallis had once performed sweeping tasks under seypi but vaguely pointed out that this was the case sometime before partition in 1947.

were likely to become even more intermittent now that some of the richer Gondals were starting to hire combine harvesters during the wheat harvest. Gondals with smaller holding didn't do so yet because combine harvesters made the chaff that they used to feed their livestock unusable, but new combine harvesters were being designed to deal with this problem.

While changes in the rural economy meant that villages were no longer the main centres of economic activity these changes didn't undermine the political and economic dominance of the Gondals. According to Herring (1983), the green revolution was a time when South Asian rural elites invested in supra-local electoral politics in order to maintain their dominance and to bolster labour-repressive social organisations of production which traditional patron–client ties no longer sustained. They also did so to get access to improved capital works, technical information and inexpensive credit. In the process their focus shifted away from the village and its inhabitants towards urban centres which were the centres of political and economic activity.

In the Punjab this was particularly true from the 1970s onwards with the rise of populist politics when broader segments of the landed classes – empowered by the green revolution and by state patronage during the military regime of Ayub Khan – started competing for access to the state and spending more time in urban centres where political power was concentrated. In order to reproduce and consolidate their class position by gaining access to state resources, they started seeking positions within elected and non-elected branches of government. As in India, it became common for landlord households to try and place members in as many different branches of government as possible. One son might join the police, another the judiciary, another might join the military, and yet another might become involved in politics. Difficult entrance exams for positions within the bureaucracy and the higher ranks of the military meant that landlords had to invest heavily in the education of their children. Often the least academically gifted son was made to care for the family farm. The overall result was that the livelihoods of landlords ceased to revolve solely around the village, just as was the case for labourers and sharecroppers.

This trend was clearly visible in Bek Sagrana, whose dominant Gondal clan (*biraderi*) belonged to the category of middle to large landholders that gained political prominence in the 1960s and 1970s.[23] Because until recently the Gondals had practiced preferential cousin marriage – which

---

[23] Hamza Alavi (1973) classifies farmers with between zero and five acres as small farmers, those with between 5 and 25 acres as medium farmers and those with over 25 acres as large farmers.

created multiple overlapping kinship ties – they formed a compact, tightly organised and localised clan which could effectively cooperate in the pursuit of economic and political power (see Alavi 1972a: 26). It was this that had given them a decisive advantage in electoral politics over the Makhdooms who were wealthy but who lacked both numbers and muscle power.

In the Bek Sagrana proper, 8 out of 13 Gondal households (out of a total of 123 households) constituted 74 per cent of the landowning population, and in nearby settlement clusters surrounding the farmhouses of the wealthiest Gondals the land distribution was even more skewed with Gondals constituting over 95 per cent of the landowning population. In the village only two Gondal households owned no land and worked as overseers (*munshis*) for wealthier Gondals. The average size of landholding among the remaining 11 Gondals in the village was of 25 acres, with the poorest Gondals owning 3 acres and the wealthiest 100 acres. In settlements beyond the village, the wealthiest Gondal in the area as a whole, owned over 400 acres and the second wealthiest owned 300 acres: both moved out of the village into newly built spacious compounds in the 1980s. All except the two poorest landowning Gondals in the village spent significant amounts of time in Sargodha where their children were being educated, but the very wealthiest lived and educated their children in Lahore.[24] They returned to the village on a weekly or fortnightly basis and only spent extended periods of time in the village during the citrus and wheat harvests as well as during elections. A number of them supplemented their agricultural incomes with government jobs in the departments of health, education and agriculture as well as by setting up small shops in town or by buying small commercial properties to let.[25] Many of them also supplemented their incomes by working as contractors (*thekedars*) building government infrastructure when the leading Gondals of the village where in power and could provide them with contracts. Some of the more disreputable Gondals supplemented their incomes by trafficking stolen buffalos, cars and even narcotics. Most recently, young Gondals in their late twenties were, in addition to obtaining government jobs, also getting jobs in banks and corporations and, since 2008, migrating to Europe and North America.

---

[24] Only the poorest Gondals in the village – two siblings who owned 3 acres each – lived permanently in the village but sent their children to live with their youngest brother who lived in rented accommodation in Sargodha.

[25] In 2005 an acre of citrus gave an income of between 60,000 and 70,000 rupees (£600–£700), meaning that a landlord with 25 acres could have a yearly income of roughly £17,000, untaxed.

Patterns of settlement and social life

Like many villages on the banks of the rivers running through the Punjab, Bek Sagrana was an old settlement (*purani abad*). Unlike the planned chaks, its square mud and brick (*kaccha*) houses and its few concrete (*pakka*) ones had been erected haphazardly, so that the narrow alleyways of the village rarely formed a straight line. The sides of these alleyways had open gutters where dark water covered with black froth flowed out of people's houses towards the village pond (*talaab*), and the alleyways themselves were strewn with litter. Behind the mud and brick walls, and the closed doors facing the alleyways, lay dusty courtyards at the end of which stood the square mud and brick houses of the majority of villagers. Only 37 out of the 118 houses that composed the village were pakka, and all of these belonged to Gondal and other zamindar households. The rest of the houses were built with bricks that were held together with mud rather than cement, while some were built purely with mud and straw. These houses generally possessed two or three rooms in a row. One of these rooms was normally used to keep the bulk of the family's most prized possessions, which often included ornately painted string cots (*charpais*) as well as copper and tin vessels and plates displayed on shelves that covered one or more walls. This room also generally contained large metal trunks where people stored the wheat that they ground and consumed as bread (*roti*) throughout the year and which they could use as currency to pay for basic foodstuffs in some village shops.[26] This was also the room that was generally given to valued guests when they came to spend the night.

Slightly beyond the village proper, there were various settlements surrounding landlords' farmhouses (*deras*). These settlements generally emerged wherever a landlord built a farmhouse around which his servants and kammis came to settle. In some instances these deras had up to two dozen kammi households. In the area around Bek Sagrana the largest dera was that of Sufi Ahmed Abbas Gondal, who owned approximately 300 acres of land. As will be shown in chapter six, Sufi Ahmed Abbas Gondal moved out of the village in the early 1980s in response to a spiritual calling urging him to set up a Sufi place of worship at a distance from the corruption and strife of the village, taking several kammi families from the village along with him. By 2004 the dera had grown to have 22 households.

---

[26] There were two small flour mills in the village where people could turn their wheat into flour. One of these mills was owned by the village blacksmiths (Tarkhans) and the other was owned by a potter (*Kumhar*) family. Both mills retained an eighth of the flour produced as their fee.

Of these, 20 were kammi households, one was the village Imam's and one was Sufi Ahmed Abbas's. Apart from Sufi Ahmed Abbas, only the village Imam's household and one Mirasi household owned any land. Within less than a mile's radius of the village there were three other such deras that had been established since the 1990s by the wealthiest Gondals, seeking to escape the factional strife then raging in the village. On these deras – as in the village – the only communal gathering places where on premises that belonged to Gondal landlords but even these places were increasingly unused since the Gondals were often away.

Other than during the cold and foggy months of December and January, people spent their time at home in their courtyards, where charpais were neatly set out side by side or in a C-shape, and in such a manner that their ends, where people lay their feet, didn't point westwards towards Mecca. In the mornings the men would sit on these charpais in a cross-legged position, and would be served their breakfast by their wives before setting off to work. Breakfast for the poor generally consisted of the warmed-up leftovers from the previous night or leftover roti with a chunk of unrefined sugar (*gur*) and, if lucky, a cup of watery tea. For those who were slightly better off breakfast might include a glass of buffalo milk; a valued source of nutrition held to be an important source of strength (*takat*) and vitality. For the wealthiest chowdris breakfast might include eggs and *parathas* fried in *desi ghee*, mango pickle (*achaar*), yoghurt, and sweet tea prepared in fresh boiled milk. They too ate on charpais placed in their courtyards, although during the devastatingly hot summer months some of them ate their breakfast in air-conditioned rooms. While the men sat on charpais, their wives and daughters (and servants in the case of the rich) squatted in a corner of the courtyard, where they prepared the fire and cooked.

During the day the women of the Gondal zamindar households, who maintained strict purdah, spent much of their time in the courtyard attending to household chores while their men were away on business. On sunny winter mornings they laid out their charpais in the sun to warm up, and when the heat mounted they retired under the shade of a tree or to the shade under the gallery. Poorer kammi women, who couldn't afford the luxury of purdah, often had to leave their houses in order to fetch wood for cooking, fodder for the livestock and kitchen provisions. Some of the poorest women also left their houses in order to do domestic work for the Gondals. Afterwards they would return home to prepare a simple lunch for their husbands and children, and a somewhat more substantial meal in the evenings.

During the hot summer months, when temperatures could reach 50 degrees centigrade, entire families would sleep in their courtyards at night. Those who could afford to do so brought out their electric fans and kept them running throughout the night. Those who had a fan but couldn't afford to pay their electricity bills stole electricity and, before sunrise, furtively pulled down their illegal electricity connections in order to avoid having to bribe the local Water and Power Development Authority (WAPDA) officer. In houses without a fan the men would sleep on the roof in the hope of catching a cool gust of wind while the women, for the sake of modesty (*hayaa*), generally stayed below in the courtyard. The wealthiest Gondals had air conditioners so, if they were in the village and not away in their town houses, they slept inside and caught colds in midsummer. They too sometimes stole electricity in order to avoid the massive bills that resulted from running air conditioners.[27] When there was too much strain on the electrical grid, which was in part caused by wealthy people running their air conditioners, blackouts and load-shedding meant that, for some hours, both the rich and the poor suffered together.

Almost all courtyards possessed a handpump (*nalka*) for drinking water and washing, with only a few households having running water. Those who could afford to do so erected walls around the handpump in order to be able to wash themselves with greater privacy. Only the wealthiest Gondals built bathrooms with hot running water inside their large houses. Many people also used their courtyards to keep chickens and tether their livestock overnight, when cattle and buffalo thieves roamed the area. For their part, wealthy landlords had cowsheds where their servants and bondsmen spent the night in order to guard the livestock from thieves. While most households possessed nalkas, very few possessed latrines (only 15 out of the 118 households in the village did so). All but two of these were Gondal households. Men whose houses didn't possess latrines either used the latrines within the compound of the Friday mosque (*jamia masjid*) in the centre of the village, or went out into the citrus orchards where they could perform their bodily functions out of sight. Because their wives and daughters weren't meant to enter the mosque they went out into the orchards in groups, either under cover of darkness for the sake of modesty, or in the early morning. Villagers told me that women went out in groups to protect each other in case wandering men, particularly local Gondal chowdris, tried to take advantage of them.

---

[27] Running a single air conditioner every night could result in electricity bills of over Rs10,000 per month. In order to avoid such bills landlords sometimes set up clandestine connections to the electrical grid or bribed the local Water and Power Development Authority (WAPDA) official to tamper with their electricity meters. The poor couldn't afford the bribes so they simply hoped not to be caught.

In the centre of the village, towering above the rest of it, were the fortress-like houses of the wealthy and politically influential Gondal chowdris. Unlike all other houses in the village, these houses had two storeys and some even had towers from which their owners could keep an eye on the activities of villagers. All of them possessed a walled courtyard where the chowdri womenfolk (*chowdranis*) households often spent their days attending to guests and chatting with the other women of the village, as well as supervising their servants sweeping the courtyard, washing clothes and cooking. Some of the older houses, built or refurbished around the 1950s and 1960s, had art deco designs plastered upon their walls and ornately carved heavy wooden doors. Their insides were cool as a result of the thick walls and the high ceilings with small vent-like windows at the top, which were opened with long wooden poles in order to release the hot air that accumulated near the ceiling. The furniture in the older houses was old and faded, and the rooms containing it had a musty smell suggesting the decay of long years.

At the time of fieldwork three out of five of these large houses had been abandoned as a result of the violent factional conflict that raged in the village throughout the 1990s (see chapter three). Thanks to the rising incomes that followed from the green revolution and from the introduction of citrus orchards as well as from involvement in politics and crime the houses' owners had been able to move out of the village and build new houses inside walled compounds on their lands within a mile of the village. When they moved they took their servants and kammis along with them. In this way three large settlements were created outside the village. As mentioned previously, these settlements were referred to as deras. The houses that the landlords built there were closer in style to the houses with imitation neoclassical features found in the housing colonies of suburban Lahore than to their former village dwellings. As noted, they sometimes had several storeys, and some had entrances that were flanked by tall white columns. Unlike the older dwellings some of these houses had large windows, which were tinted green or black in order for the women to be able to look out without being seen from the outside. Their walls were thinner than those of traditional havelis and they therefore quickly heated up if an air conditioner wasn't kept running. Rather than dusty paved courtyards at their centre, these houses had large walled gardens at their back where the women, who now spent most of their time in Lahore, could spend their days in seclusion. The walls around the perimeter of these gardens, which in some cases covered well over an acre of land, had high walls studded with shards of glass in order to prevent thieves and enemies from entering.

The kammis and servants who accompanied the landlords when they moved out of the village settled around the outside perimeter of the landlords' walled compounds. The green lawns and the imitation neoclassical buildings inside the walled compounds contrasted sharply with the traditional flat-roofed mud dwellings that clustered on the outside. The physical distance of these compounds from the village, as well as the stylistic distance between them and more traditional dwellings, reflected the widening social and economic gulf that increasingly separated the Gondals whose fortunes had soared from the majority of villagers. Whereas the previous generation of village leaders, who were now either very old or dead, had lived most of their lives in the village, their sons, and particularly their grandsons, increasingly spent their time in cities such as Lahore or Sargodha. One important factor that had facilitated their move to the cities had been the introduction of citrus orchards which only required minimal supervision from their owners.[28] The older generation had gone to school in the village, where they had studied and played with poorer villagers whom they had come to know intimately. Later on in life they had assumed village leadership and spent their days attending to village problems in the village men's house (*darra*) next to the Jamia Masjid. Nostalgic old villagers recalled how the village leaders, who at the time weren't divided by factional rivalries, had presided over men assembled at the darra and exchanged gossip and jokes with them on a daily basis.

Since the early 1990s the village darra had rarely been used as a place where chowdris and other villagers gathered in the evenings. The only people who used it were some disreputable Gondal siblings who owned 10 acres of land and who gambled and drank there. They were often abusive towards kammis that walked past. Sometimes they would imperiously send them on errands and reward them with abuse and denigrating jokes

---

[28] In a similar manner the Anavil Brahmin landlords described by Breman (1974) had been able to move to the cities as a result of the introduction of mango orchards. However, because the economic and political fortunes of the Gondals were closely linked to their involvement in politics they needed to maintain control over their constituencies. This meant that if they became excessively urbanised and lost touch with their rural constituents, they risked losing their political clout. Thus unlike Breman's Anavils, whose relationship with labourers had become largely contractual, the relationship between Gondals and labourers and other villagers continued to incorporate political, economic and religious factors. Whereas for the Anavils having large numbers of dependents no longer played an important political function, the Gondals still needed dependants to settle their feuds and to win elections. One of the aims of this work is to explain why political power remained central in guaranteeing the fortunes of Gondal and other landlords.

Plate 1.1 View of Bek Sagrana with low-lying mud houses in the foreground and towering Gondal houses in the background.

Source: All photographs are by the author.

Plate 1.2 Abandoned Gondal house in the centre of Bek Sagrana.

when they returned. On one occasion I witnessed them making fun of a kammi by mockingly referring to him with the honorific term of address 'Chowdri Ji', a term used to refer to zamindars, because he was wearing brand new clothes and therefore putting himself above what they considered to be his true station in life. The kammi in question later told me that they were abusive towards him because it bothered them to see a lowly villager dressing the same way they did. He said that they would rather see him walking around in rags because they wanted people like him to remain poor and dependent so that they could continue to boss them around.

When the sons of those who had once presided at the village darra now attended to villagers' needs they did so at their own deras. Even then, however, their availability was limited since they were often away in Lahore where their children were being educated at prestigious English language schools and universities. Some of their children even ended up going to universities in Canada and in the United Kingdom. As a result of their upward mobility the distance between leading landlords and their village had grown progressively over the generations. Thus, whereas the landlords who were now over 70 had been to school in the village, their sons had gone away to schools, and in some cases to universities, in Sargodha while their grandchildren had been educated in Lahore. The relative proximity of Sargodha to the village, and its role as the principal administrative and commercial centre for villagers, meant that the middle generation had been able to maintain regular contact with the village. They had returned to the village on weekends and had frequently met villagers in Sargodha when these went there for shopping or for administrative purposes. In Sargodha they were able to forge friendships as well as enmities with the young members of the district's future political elite. Later on in life when they entered politics, these ties and enmities would play an important role in the formation of local political coalitions (see chapter two). As members of this middle generation rose to power at district, provincial and national levels they were able to amass large fortunes and to engage in a process of upward social mobility.

Members of the six leading Gondal families now increasingly bought or rented property in Lahore where the provincial and much of the national elite resided. Here they were able to rub shoulders with senior civil servants, judges, politicians, rural magnates, and industrialists. Members of the old aristocratic landlord families, such as the Noons who lived in Lahore and who no longer wanted to be involved in politics, regarded them as upstarts. Nevertheless, because the Gondals were able to get some of their children into elite schools (such as Aitchison College),[29] they started associating and making friends with the sons of the wealthiest and most powerful people in the country and shedding the more rustic social veneer of their parents in the process. Whereas the middle generation had rubbed shoulders with the district and provincial elite, the third generation did so with the provincial and national elite. The result was that this third generation harboured far greater ambitions than their parents had. They hoped to gain high ranking political office or jobs in multinational companies, banks and law firms, and they saw the village as a backward and violent place that they preferred to avoid. When they went there they felt out of place. They only came to the village three or four times a year and on these occasions they often remained inside playing computer games, watching television and smoking marijuana or drinking alcohol, which had been secretly provided by their servants. I soon realised that after having spent only a couple of months there I knew more of its inhabitants by name than they did. Like the Makhdooms before them, this third generation of Gondals was losing touch with the village and its ways.

Besides the most prominent Gondal households, the other eight Gondal households of the village generally owned between 5 and 20 acres of land and remained somewhat more rooted in the village than the leading families. There were a number of such Gondal landlords who also lived in deras not far from the village. These deras were smaller than those of the leading Gondals and generally had no more than two or three mud houses for kammis and servants attached to the main house. Villagers often referred to them as 'motorcycle-riding chowdris' (*motorcycle wale* chowdri) because unlike the wealthier Gondals who drove around in cars and pickup trucks, they drove back and forth to Sargodha and around the village on motorcycles. Moreover unlike the leading Gondals who employed people to help manage their farms, most of these Gondals were directly involved in the management of their farms. They were also more involved in everyday village life than members of the leading families who operated at higher levels of politics. When villagers needed to resolve a dispute or required patronage with a government institution, these Gondals were more readily accessible

---

[29] The college was founded in 1886 by the then governor of the Punjab Sir Charles Umpherston Aitchison. Its establishment coincided with the beginning of canal colonisation and served to educate the scions of the leading landed families of the Punjab. Over the years the college produced a large number of prime ministers, chief ministers, senators and judges and continues to do so.

than were the members of the leading families. As will be shown in chapter five, during elections they played an important role as brokers between the more powerful Gondals and poorer villagers.

Within this category of Gondals some of the more hard-working and academically gifted youths were studying in Sargodha and some were preparing for provincial and national civil service exams. Others studied law and one, who was middle-aged, had become a doctor. In other cases they had acquired civil service jobs through patronage or, as locals put it, through the 'recommendation' (*sifarish*) of one of the more influential village Gondals. Still others became involved in politics at the union council level, the lowest tier of elected government, and acted as brokers for the higher-ranking Gondal politicians in the village. Since my departure in 2006 three young men belonging to this class migrated to Britain, Switzerland and France on student visas in order to work in shops, restaurants and minicab companies. All of them placed high hopes into migration and aimed to return to Pakistan with money to invest into land, business or even into politics.

Obtaining new sources of income was seen as a necessity by those of them who realised that they would otherwise end their lives poorer than when they had been born, since the subdivision of land across generations meant that they would inevitably have less land than their fathers. If they did nothing to compensate for this then they and their children might in some cases end up dangerously close to poverty. However, not all of them pursued further education, and less-academically gifted young men generally ended up taking care of their lands. Other young men from within this category ended their studies at a young age and became involved in crime. Such was the case with the young Gondals, referred to previously, who had occupied the village darra. They adopted the swagger and the twirled moustache with the unshaven look that typified the village Tough (*badmash*), and they spent their nights playing cards, drinking home brews and smoking hash. One notorious set of four brothers, whose rise and fall will be documented in chapter three, became involved in bootlegging, petty heroin trading and the smuggling of stolen cars. In the process, one of the four became a heroin addict himself. Young men such as these frequently attracted the attention of more powerful Gondals, who made good use of village tough for political and economic ends. Powerful landlords offered them protection from the police in exchange for a share in their businesses. They also used them to fight opponents and to intimidate and harass unruly villagers, particularly during election times.

By virtue of the fact that these less prominent Gondals belonged to the dominant lineage and shared kinship ties through marriage with the most powerful Gondals, their power and status was superior to that of members of other landholding biraderis (such as Ranjhas, Lurkas and Rajputs). Their kinship ties with the powerful village Gondals made it easier for them to obtain the patronage necessary to gain certain government jobs, to set up a business and to obtain favourable outcomes in legal disputes. The result was that even relatively poor Gondals had more influence in the village, and were more likely to have government jobs, than were members of relatively well-off non-Gondal zamindar households. The latter had little influence in village matters or in local politics, and they dedicated themselves to running their farms and setting up small businesses. Some of them opened shops in the village or in the nearby market town of Chowki Bhagat, others worked as middlemen during the citrus harvest, and others were involved in the timber business. This suggests that membership of the dominant biraderi made a significant difference to class positioning because it placed people in a more advantageous position vis-à-vis the state.[30]

---

[30] This finding coincides with Chakravarti's claim that membership of the dominant *Bhumiar* caste in a village in Bihar played a crucial role in 'determining access to the means of production, control over the labour process, and forms of exploitation' (Chakravarti 2001: 106) and consolidated the class position of its members.

2
DEBT AND BONDAGE

From patronage to exploitation

Marguerite Robinson's ethnography on local politics in Andhra Pradesh (1987) illustrates how poverty, wealth and power are linked in a variety of intricate ways. Her work shows how Reddy landlords traditionally maintained their grip over village 'vote banks' through debt-relations with members of lower castes. It was, according to her analysis, thanks to the poverty and powerlessness of the majority that a minority of landlords were rich and powerful. She illustrates how the Reddy's ability to deliver vote banks to politicians was undercut as they were removed from their posts as village leaders through government intervention, through land ceilings and through the delivery of government credit to villagers. This chapter illustrates how in contemporary Pakistan – more than three decades after the Reddy's control over debt-relations had been undercut by the government – landlords' grip over vote banks was still partly due to their control over credit relations. By providing credit to labourers, landlords assured themselves a cheap and regular supply of labour as well as the votes of entire households. I argue that they were able to ensure the servicing of debts through labour by exerting pressure and threats on entire households. Following Brass (1999), I argue that an important consequence of this was to displace potential class conflict onto the kinship group.[1]

---

[1] While I follow Brass on this point, my broader argument is closer to that of authors who argue that contemporary forms of debt-bondage reflect the proletarianisation of the workforce. In the Indian context authors such as Patnaik and Dingawey (1985) and Brass (1999) argue that forms of labour attachment involving coercion are still widespread and deprive labourers of proletarian status. The result of this is that they cannot challenge their own subordination. On the other hand writers such as Rudra (1994), Prakash (1990) and Breman (1974, 1993) argue that rural labourers in India are increasingly proletarians in a capitalist system. This means

In its World Labour Report of 1993, the International Labour Organisation (ILO) described the problem of bonded labour in Pakistan as among the worst in the world (ILO 1993). According to a 1994 estimate by the Human Rights Commission of Pakistan (HRCP), around 20 million people worked as bonded labourers in agriculture, brick kilns, fisheries, construction, and domestic service (HRCP 1995: 120). Reports suggest that bonded labour is most widespread in agriculture and most coercive in areas of southern Punjab and interior Sindh, where land distribution is particularly unequal. Landlords in these areas are known to have maintained private jails where labourers were kept locked up and guarded by armed men at night, and sent to labour in the fields by day. Female labourers were frequently assaulted by landlords and their strongmen as well as by the police who were complicit in the maintenance of these jails. In 1991, in a widely publicised case, the army raided the private jail of a major landlord in Sindh and released 295 labourers.[2]

In the central Punjab district of Sargodha there were no private jails and I didn't hear of anyone operating any in the recent past. Nevertheless debt-bondage, whereby labourers might spend a lifetime servicing a debt that could occasionally pass on to the next generation, was widespread. Although landlords didn't jail their labourers, they had other means of coercion at their disposal in order to ensure the repayment of loans and the effective exploitation of labour. Foremost among these was their ability to exert pressure on the immediate kin of indebted labourers when the latter shirked, absconded or ran away. When this happened, landlords could use the threats of eviction, denial of access to firewood and fodder,

---

that the relationship between them and their employers is increasingly one which is devoid of both direct coercion and patronage. For Breman the assertion of the continuity of historical forms of bonded labour in modern day Gujarat ignores the 'materialistic and ideological changes that have occurred between landowners and landless in the countryside in South Gujarat' (Breman 1993: 311). He argues that the onset of commercial agriculture and the expansion of the Indian state have largely eliminated the elements of both patronage and direct coercion that used to characterise the relationship between Anavil Brahmin landlords and their Dubla labourers and servants previously. While these changes partially liberated the landless from oppressive relations of domination, they also deprived them of customary – patronage-based – subsistence guarantees and left them vulnerable to the vagaries of the labour market.

[2] See Breman and Lieten (2002) for an account of bonded labour in rural Sindh. Their work focuses principally on Hindu Hari sharecroppers in Sindh where sharecropping remains more widespread than in the Punjab.

fines and physical harassment to make his children, or siblings – if they still formed part of the same extended household as the labourer – service the debt in his place. If the labourer had no able-bodied men in his household, then the possibility that his wife and children might get evicted or harassed if he absconded was generally enough to deter him from doing so. Since the families of bonded labourers knew that the burden of servicing a debt might fall on them, they helped ensure that landlords got their money back by helping, and sometimes pressuring, indebted kin to repay their debts. The last section of this book will examine how this worked.

Although debt-bondage was formally banned under Pakistani law, it operated with the tacit as well as active cooperation of the police and other authorities because the landed elite had captured the state and used it to further its own interests regardless of the law. This meant that landless labourers couldn't rely on the state to redress their grievances. They regarded the state for what it was: an instrument of landlord domination.

The mechanisation of agriculture starting in the 1960s, the decline of village crafts, the shift away from sharecropping, and the introduction of citrus orchards, all significantly eroded traditional patron–client ties. Tractors and citrus orchards reduced the need for a permanently employed workforce of sharecroppers and labourers. Mechanisation helped reduce production costs and the citrus orchards allowed landlords in Sargodha district to get away with minimal supervisory work and permitted those who could afford it to move to town. The replacement of village crafts by industrially produced goods also did away with their need for village artisans such as potters, cobblers and weavers. Using data compiled by Khan (2006) Table 2.1 displays trends in the use of wage labour in the Punjab as a whole, showing a decline of permanent hired labour from 7 to 3.4 per cent and a rise in casual labour from 30 to 44 per cent.

Table 2.1 Types of labour employed in agricultural households (%)

	1972	1980	1990	2000
Family workers as percentage of family members	55	60	29	37
Permanent hired labour	7	4	2	3.4
Casual labour	30	45	50	44

Source: Khan (2006).

Moreover although Punjabi landlords often used armed toughs to fight and intimidate rivals this didn't make them 'feudal lords' who derived power and prestige from the number of fighting men under their control – like the Khans described by Barth (1959) and the Anavil Brahmins in historical Gujarat described by Breman (1993). Modern day Punjabi landlords derived most of their power from the control of state institutions. To achieve such control they were more likely to invest the surplus from their farms into private educations, so that their children could obtain civil service jobs, than into creating a loyal client base to fight rivals. Moreover status was now more likely to be associated with consumer goods than with large retinues of servants.

The overall result was that, as in Breman's Gujarat (1993) and Jodhka's Haryana (1996), there had been a significant erosion of the ideology of patronage and loyalty.[3] Landlords no longer sat in the men's house gossiping with villagers and resolving their disputes. Moreover most of them no longer took notice of key life cycle celebrations in villagers' lives. Their relationships with labourers had become contractual and restricted to key moments in the agricultural cycle. Moreover, because landlords, increasingly concerned with costs of production and maximising profits to fund new forms of consumption, overworked and exploited their few attached servants, not many people wanted to become their permanent employees. Some did, but only in exchange for loans, and even this tended to be a last resort. Unlike in traditional settings where bonded labourers weren't expected to ever repay their debts, and where labourers themselves might not want to do so because they preferred being attached to a landlord than free and unprotected – such as historical Gujarat described by Breman (ibid.) – labourers now sought to escape servitude by repaying their loans as quickly as possible. Their ambition was now to obtain a secure formal sector job either in the public or in the private sector. Although many succeeded in escaping servitude, few ever achieved the dream of a formal sector job and most were condemned to a life of itinerant wage labour.

This chapter documents the increasingly exploitative and antagonistic relationship that arose out of the decline of traditional patron–client ties. I argue that the exploitative conditions documented in the book, as well as the even more extreme ones reported by the Human Rights Watch, are possible because despite the proletarianisation of the workforce, military and or political interventions forestalled the emergence of popular political movements aiming to secure basic rights for labourers. Without genuine

---

[3] Scott (1972) has discussed the erosion of the ideology and patronage in the context of Malaysia.

political representation, Pakistan's rural poor have been unable to secure the right to either a minimum wage or to secure housing rights (see Gazdar and Bux Mallah 2011), let alone to health care, education and justice. Instead I argue that landlords have been able to use their stranglehold over the state in order to intensify their exploitation of the workforce.

Wage labour versus attached labour

Becoming a farm or household servant attached to one of the Gondal chowdris was not an attractive option to most people. Nevertheless the seasonal variations in the availability of daily wage labour as well as the need for credit, to cover things such as large wedding or medical expenses, and other forms of patronage meant that servitude was an option that was resorted to frequently. It was well known that working for a chowdri was both physically and mentally exhausting. Servants were poorly paid and had to be at the constant beck and call of their masters who, as the servants often claimed, treated them like animals.

The highly seasonal demand for agricultural labour was largely the result of the widespread introduction of tangerine and orange orchards that began in the 1970s. This had also played an important role in reducing the area of land that the chowdris dedicated to sharecropping and causing many redundant sharecroppers to turn to wage labour as a result. The chowdris had introduced citrus because it was both more profitable and less labour-intensive than other forms of cultivation, which had previously included sugarcane, wheat, cotton, maize, fodder crops, rice, and vegetables. This allowed the chowdris to spend less time in the village overseeing agricultural tasks and village affairs and more time in the cities where their children were now being educated.

The greatest demand for labour in the citrus orchards occurred during the harvest season over the three months between the end of December and the end of March. A contractor (*wapari*) who purchased the fruit based on roughly estimated yields while it was still on the trees was in charge of organising the harvest. This included recruiting and organising labourers as well as transporting and selling the fruit. Depending on the size of the orchard and on the speed with which the contractor wanted to complete the harvest, citrus harvesting teams could comprise as many as 40 people. Such teams comprised fruit pickers who climbed the trees, people to carry away the fruit in large wicker baskets and load the truck(s) and sometimes a group of people to sort the fruit according to grades of quality and box it. At this time of year it was common for various contractors and their teams of labourers to be working simultaneously in the orchards of different chowdris.

The chowdris had nothing to do with the process other than finding the contractor that offered them the best price for his services. During this time, therefore, labourers dealt directly with the contractors rather than with the chowdris. Fruit pickers earned between Rs 100 and Rs 120 per day making it possible for them to earn around Rs 2,500 per month if they worked 25 days each month over the three months of the harvest. This was significantly more than what attached farm or domestic servants earned in monthly wages, and it was also more than what daily wage labourers could expect to earn throughout the rest of the year with the exception of the wheat harvesting season.

Outside of the harvest period, however, the demand for labour in the citrus orchards was relatively low and the chowdris or their managers (munshis) organised and recruited labour for the few tasks at hand without the assistance of a contractor. The orchards had to be irrigated throughout the year and in particular during the flowering of the trees in the spring, a task that was generally carried out by a single labourer, usually an attached farm servant. The orchards also had to be fertilised once a year and sprayed with pesticides about six times a year. These tasks were also frequently carried out by a farm servant and a few hired wage labourers who were paid a standard wage of Rs 80 per day. In addition, during the summer months, some labourers were hired to trim the trees and remove any dead branches that had not already been removed for firewood.

The second greatest demand for agricultural labour occurred shortly after the end of the citrus harvest for the wheat harvest that started around the middle of April. Although citrus provided the bulk of the Gondal chowdris' agricultural income they all dedicated some of their land to wheat and fodder cultivation. It was possible to cultivate both of these crops in the citrus orchards when the trees were still relatively small and didn't stop the sunlight from reaching the crops. For the most part, however, the chowdris saved some land solely for the cultivation of these crops. Wheat in the form of unleavened bread (roti) was the basic daily staple of ordinary villagers and chowdris alike and all households stored wheat to last them for the entire year if they could afford to do so. Additionally the chowdris kept an important amount of it in order to pay their servants and the kammis who occasionally provided services to them and were paid fixed amounts of wheat on a yearly basis.

During the wheat harvest plots of land were allotted by the chowdris to teams of labourers of up to five people who were frequently, but not necessarily, close kin belonging to a single household. The work was known to be physically exhausting so only the able-bodied took part in it, but it was also considered to be the most important and profitable work of the year. Teams of labourers were paid three maunds of wheat per acre harvested meaning that at a value of Rs 400 per maund a team of five labourers could make Rs 250 per head if they harvested an acre a day. Although wheat could be sold and directly converted into cash most people stored it for consumption throughout the year and only occasionally sold some whenever they needed money. However, work opportunities for labourers during the wheat harvest were likely to become scarcer now that some of the richer Gondals with larger fields were starting to hire combine harvesters during the wheat harvest. Gondals with smaller holdings didn't do so yet because combine harvesters made the chaff that they used to feed their livestock unusable, but new combine harvesters were being designed to deal with this problem.

Outside of the wheat and citrus harvests there was no time of the year when almost all of the villagers were working in the fields. People took up wage labour both in and out of agriculture and frequently found employment in construction and sometimes in factories in nearby industrial centres like Sialkot and Faisalabad. However, most of the labourers that I interviewed only took up industrial employment for a few months at a time and at most for a year or two. Although wages in industrial jobs were higher than those in agriculture living expenses in town were significantly higher than in the village where accommodation was free and food was less expensive. Also, industrial labourers had to live far from their families, who could not easily accompany them because of the cost of accommodation, and they were therefore frequently homesick. And, while young and unmarried men were free to go away, labourers who were the heads of independent households were reluctant to leave their wives and children to fend for themselves in the village.

Labourers from the various artisan clans (*biraderis*)[4] as well as displaced agricultural tenants told me that the main benefit of free wage labour was that it allowed them to take advantage of the peaks in the agricultural cycle during the wheat and citrus harvests when daily wages were at their highest. Another benefit was that the economic compulsions of wage labour gave them a relatively greater degree of freedom than the extra-economic forms of compulsion that they were subject to if and when they became attached servants with a chowdri. Labourers could now find work with employers as far afield as Islamabad and their livelihoods no longer solely depended upon the Gondal chowdris of Bek Sagrana (who in any case spent each time more of their time living in cities). Both chowdris and kammis told me that as a consequence villagers were relatively freer from the influence of the Gondal chowdris than they had been in the past.

---

[4] A birderi is literally a 'brotherhood' and refers to members of an extended patrilineal clan.

On the other hand, the disadvantage that labourers saw in free wage labour was that it could be difficult to find work on a daily basis outside of the peaks in demand between December and May. Free wage labourers also faced a significant degree of uncertainty as to whether and when employers with whom they didn't have longstanding relationships would actually pay them. In addition to this free wage labourers faced greater difficulties than attached farm and domestic servants in obtaining patronage and loans, both small consumption loans as well as larger loans to cover wedding and medical expenses. Both the uncertainty of free wage labour and the need for patronage meant that many free wage labourers opted to become attached servants with the chowdris, sometimes out of necessity and at other times simply for the sake of security, despite lower wages and working conditions that were known to be exploitative and abusive. Besides allowing them to take both small consumption and larger loans, labourers also took up employment with the Gondal chowdris because it was supposed to entitle them to other forms of patronage including support in case of a dispute, mediation with state institutions and even help in retrieving an eloped wife or daughter.

On their side, part of the reason that the chowdris extended such forms of patronage and gave labourers large loans was that they thereby ensured themselves a cheap and permanent supply of labour. This was important to them because in a great economic circularity permanent and reliable labour wasn't readily available due to the reluctance of many labourers to work for them. Chowdris frequently complained that nowadays the only way to obtain labour was by giving loans and patronage. Many chowdris blamed this on the fact that the kammis had inflated job expectations as a result of a combination of education, television and easier access to urban centres. Until the time when the debt was fully repaid chowdris guaranteed themselves cheap labour at their beck and call by extending loans. Because of the Islamic prohibition on interest these loans were interest free. Nevertheless, as will be shown below, landlords still profited because extending the loans allowed them to reduce their workers' wages. Debts could be repaid after weeks or after an entire generation depending upon both the size of the loan and upon whether the labourer had able-bodied siblings that could help him in the process of repaying. For the chowdris having servants was also part of a display of status that demonstrated power and influence. Unlike in modern day Gujarat (Breman 1993) having dependants in the Pakistani Punjab continues to play an important political function for the chowdris because servants, and especially gunmen, are frequently used to settle disputes with enemies (*dushman*) and because dependants more generally are mobilised for their votes during elections.

The sorts of labourers that the chowdris generally required and to whom therefore they saw particular advantage in extending loans and other forms of patronage were farm servants, domestic servants and drivers as well as gunmen. Depending upon the chowdri who employed them it was not uncommon for attached servants to end up having a combination of these duties. The most usual combination was of domestic and farm work, despite generally having been hired for either one or the other. Only in instances where they were hired by village factional leaders involved in politics was it usual for labourers to become gunmen, and only the wealthiest chowdris were likely to hire drivers. Farm work usually involved feeding and milking the livestock, irrigating the orchards and the crops, sowing crops, and driving the tractor. The domestic tasks that farm servants ended up having to attend to often included serving food and tea to the chowdris and their guests, cleaning, and being constantly sent on errands to fetch and buy things, frequently well into the night. Even though they might have their own house in the village, servants were frequently asked to stay the night in order to guard the livestock from roaming cattle thieves and to be on hand if the need for their services arose. Additionally, attached servants had to work throughout the year; if they ever wanted a day off it was generally the case that they personally had to provide a replacement to work during their absence.

For all their labour, farm and domestic servants rarely earned more than Rs 1,500 per month plus eight maunds of wheat per year, two meals and a cup of tea a day, and at least one set of clothes per year on Eid. In total this amounted to roughly Rs 18,000 per year or Rs 70 per day, which was less than the lowest wages of the agricultural cycle (usually around Rs 80 per day) and considerably less than peak daily agricultural wages earned during the citrus and wheat harvests. However if the servant happened to have a large debt his wages were often significantly lower at around Rs 10,000 per year. It therefore appears that the reason that the chowdris extended loans to servants was because it allowed them to reduce their wages and thereby assure themselves a cheap and permanent supply of labour.

Child servants

The cheapest source of labour available to the chowdris was children as young as five years old. Poor households sometimes sent their children to work in the village or town homes of the chowdris. In exchange for their child parents either obtained some form of patronage, which might include a loan, or payments of around Rs 6,000 per year. This practice was justified by the chowdris with the argument that they were doing the children and their families a great favour by hiring them because chowdris themselves would be able to provide for, protect and educate the children better than their parents would. One chowdri told me that the 11-year-old girl that worked in his house in Lahore was there for the protection of her modesty (*hayaa*) because there were no adult males in her family to keep the young men of the village at bay. In most cases the chowdris told me that they took on such children in order for them to learn how to read the Holy Qur'an as well as how to read and write Urdu. However, for most villagers it went without saying that the children weren't really gaining an education. Villagers knew the children were generally unhappy and homesick, and that they spent most of their time wearily washing dishes, carrying trays, sweeping floors, and running errands. Those children who ended up in Lahore or in Sargodha often didn't see their families for months on end. In fact many chowdris purposely avoided taking the children back to the village to visit their families because they claimed that these visits subsequently made them more homesick and less willing to work.

The mindless drudgery of the children's work was clearly illustrated by the case of one seven-year-old Mussalli boy who worked for the wealthiest chowdri in the village. The child's father had abandoned his family and left its members to fend for themselves. The child's mother had started to work as a servant for a Gondal chowdri but had been unable to make ends meet and was compelled to take several loans from her master. Indebted and still short of money, she eventually accepted Sufi Ahmed Abbas's offer to employ her son. Sufi Ahmed Abbas offered to pay her Rs 400 per month for her son's labour and he also promised to feed, clothe and educate him. Even though her son would be required to spend most of his days and nights at Sufi Ahmed Abbas's farmhouse, the fact that he would be nearby meant that she could occasionally see him in the little free time that she was granted by her own master. Moreover she consoled herself with the idea that Sufi Ahmed Abbas and his wife would be able to give him a better education than she ever could.

The reality however was that her son spent little time studying. His working day started at around five in the morning when he was often sent laden with heavy breakfast trays, which he clearly struggled to carry, to the men's houses where guests were staying. The same tasks were often repeated at lunch and dinner times. Between these periods he was largely occupied carrying and fetching things for his masters. Like other chowdris, his masters preferred being served than doing things for themselves, which meant that there was almost no limit to the number of times he might be sent to fetch a glass of water, a cup of tea or a packet of cigarettes. I witnessed several instances when he would be called from another room in order to bring a glass of water that was only a couple of metres away from his master. He also spent so much time running around communicating messages between his master and different labourers that villagers came to refer to him affectionately as 'Nokia' after the brand of mobile phone.

Needless to say the repetitive and mindless nature of the tasks that he was set meant that he often appeared weary and bored. As a result he was often careless and his masters frequently reprimanded him for it. By evening time he was generally exhausted and I often saw him fall asleep on the spot at the men's house while guests were eating. Even though his mother lived close by he didn't get so many opportunities to see her, and like other children in his situation he was often homesick. There were even fewer occasions when he could see his elder sister, who had also been placed with a family as a servant shortly after he had. For her part his sister had even fewer chances to see her mother since she lived in town. However her situation was mitigated by the fact that she lived with a pious middle class family who treated her with affection and appeared to be genuinely concerned with her education. Nevertheless, despite being treated relatively well, she remained a servant and was not the equal of her masters' children. While the other children went to school she spent her days working and studying at home, and while the other children played she often had to help out with domestic chores.

Attitudes of class difference and discrimination towards the servant children were passed from the chowdris and their wives to their own children. Frequently the children of the households were told that the child servants were *junglee* meaning wild and ill-mannered. As a result the children of the chowdris quickly came to look down on the children who worked in their homes. On one occasion an eight-year-old chowdrani, who appeared to be piqued by the fact that I was giving more attention to the young boy servant than to her, told me that the boy servant was ill-mannered, dirty, lazy, and stupid and went on to explain that such was true of Mussallis generally. In addition, chowdris rarely reprimanded their children if they struck a child servant in a quarrel for example, but they were quick to reprimand a child servant if he or she had hit their child under similar circumstances. As a result young masters quickly learned that they were in a position of power and knew that they could get away with treating their servants badly. For their part, child servants often grew resentful of their masters and vented their feelings in a variety of ways. I once caught the young boy who was described by his young mistress as ill-mannered and stupid, aggressively slapping and pinching the two-year-old child of his master after being left alone with it. Given the way that he was treated around the house and his generally unhappy appearance, it is not unreasonable to assume that he was venting feelings of anger and discontent upon the only member of the household who was not in a position of power over him. Others vented their discontent in different ways. One 12-year-old boy told me about how when he had been a servant as a small child he would purposely run up large phone bills on his master's phone by calling a friend of his in town. The same boy told me that when in the early mornings his masters would call him he pretended to remain asleep in order to waste time. He also told me that on the few occasions when his masters brought him to stay with his family in the village for a few days he would go into hiding on the day that he was due to go back to Lahore.

It is clear from the evidence that even in the best cases servant children had a hard time. Adults who had been child servants almost unanimously told me that they didn't wish their children to go through the same experiences that they had despite the bonds of mutual understanding that almost inevitably developed between them and their masters. Those who spent their entire childhoods and in some cases their entire adult lives as servants were effectively inferior members of their masters' families. As family members of a sort, long-serving child servants developed some emotional attachment with the families but these ties were rarely unqualified feelings of affection and were most often strongly tinged with resentment. Such mixed feelings probably arose in large measure from the fact that while their masters inevitably took on the role of parental figures for them on one hand, as parental figures their masters were often demanding and unloving on the other. The fact that masters were parental figures for their child servants was underlined linguistically when servants who had been raised by their masters compared them to fathers and mothers (*maa aur baap*). It was also underlined by the fact that child servants often acquired the habits and mannerisms of their masters. In fact, villagers often commented that the behaviour of close servants resembled that of their masters.

The most striking example of this in Bek Sagrana was provided by a close servant of Chowdri Mazhar Ali's younger brother Chowdri Arif. The servant, named Bashir, came from a poor Machi family and had worked with Chowdri Arif and his family between the ages of 8 and 28. Even once he had ceased being formally employed by Chowdri Arif he maintained a relationship with him. He often visited Chowdri Arif in his farmhouse and performed odd jobs for him. What was striking about Bashir was the degree to which Chowdri Arif's influence upon him as a parental role model was visible. Bashir had inherited many of the marked mannerisms of his master and could easily have been taken for his son. The resemblance between master and servant was such that villagers jokingly referred to Bashir as Chowdri Arif. Bashir explained to me that it was inevitable that he should have come to resemble his master after serving him day and night for 20 years. He had shared the intimate daily routines of Chowdri Arif and his family and knew all of the family's quarrels, vices and secrets, including the many secrets that members of Chowdri Arif's family kept from each other. Such shared secrets obviously added to the intimacy between Bashir and his masters, but despite this intimacy and some material benefits Bashir felt that his time spent as a servant had not been worth it. This was the case even though Bashir could approach his masters for patronage and that Chowdri Arif had helped him to buy a rickshaw after his term of employment had ended. Bashir claimed that Chowdri Arif had tricked him into becoming his servant when he was eight by promising to teach him how to drive. Had it not been for this trick Bashir claims that he would not have voluntarily left his family. Although he eventually did learn how to drive he spent most of his childhood doing tiresome household chores. He complained that his masters had often slapped him and that Chowdri Arif frequently punished him in the manner of schoolmasters by making him squat down and hold his ears with his arms under his legs (*kaan pakarna*). He also complained that Chowdri Arif, who was a notoriously stingy man, had generally made sure that his wife served him only leftovers of food and very little meat.

Another villager who regretted his childhood as a servant was Bilal Mirasi. He was frequently singled out by villagers as an example of the psychological damage suffered by child servants. Bilal Mirasi, or Ballu as he was nicknamed, had started to work for Chowdri Mahmood Abbas at the age of five following his father's death. He continued to work for him until he was 20. Afterward, like Bashir, he had maintained contact with his masters and occasionally worked for them. For example he would sometimes accompany Sufi Ahmed Abbas's younger son when he travelled to visit friends. Like other child servants he had performed mindless work for long hours throughout his childhood. However, unlike Bashir, he had acquired few skills that might help him find employment in the wider world. While Bashir had learned how to read and drive, Bilal had learned nothing and remained illiterate.

Villagers commented that because Bilal had been treated as if he had no feelings, like an animal (*janvaar*), he had grown up to be useless as well as mentally damaged (*pagal*). They claimed that it was because of this that Bilal was unable to work consistently and spent his days smoking marijuana. As a result he was unable to provide for his children, who went around dressed in rags. His reputedly peculiar sense of logic was also used as evidence of the damage that had been done to him. People additionally observed that the damaging hardship that he had endured as a child wasn't even mitigated later by any significant form of patronage from his former masters. As evidence, villagers cited the fact that Bilal was one of the poorest men living around the farmhouse (*dera*) of Sufi Ahmed Abbas. Bilal lived in a mud hut with a single bedroom that had almost no furniture. The hut remained unconnected to the electrical grid because Bilal couldn't afford the bills. His former masters rarely offered him assistance and showed little concern for his children, who ran around naked or in rags with faces caked in mud and mucus. Once when Bilal's eldest son suffered a severe electric shock from a faulty fan (*pankah*) Sufi Ahmed Abbas offered no help in getting the child to a doctor. Instead it was the village mullah, who lived near Bilal but didn't possess even a fraction of the wealth and influence of Sufi Ahmed Abbas, who took Bilal's son to the local hospital on his motorcycle.

This evidence and more made villagers certain that the costs of being a child servant with a chowdri outweighed any of the benefits. Parents who sent their children away to work as servants were often embarrassed about it and justified it to themselves and others by invoking necessity or the need for their children to get an education. Many people told me that they would only send their child to work with a chowdri as a last resort because of economic hardship or because a chowdri had forced them to do so. People's desire for their children not to become servants of the chowdris was clearly illustrated when Dr Muzaffar Abbas, Sufi Ahmed Abbas's eldest son, asked his Mussalli driver to send one of his sons to work for him as a servant in Lahore. The driver implored his master not to ask this of him, arguing that it would break his wife's heart to see any of her children go. Later the Mussalli driver decided that his son would be much better off if he worked as an apprentice with a mechanic in the nearby market town. He argued that in this way he would actually learn a skill other than carrying serving trays. After the driver succeed in sending his son to become an apprentice many of the other kammis who lived around Sufi Ahmed Abbas's farmhouse decided to follow his example with their own children for the same reasons.

Servicing debts

The extending of loans by chowdris to kammis was discussed earlier, and it was noted that this benefited the chowdri by ensuring him of reduced-cost labour over a long period. Various factors worked together to keep this situation stable and ensure that servants remained bound to their debt obligations towards the chowdris. To begin with servants who had debts couldn't simply run away for the simple reason that they usually had no place to go. Servants had few places outside their home village they could run to and be beyond the reach of the chowdris. If they ran away from Bek Sagrana to nearby villages, or even to the nearby town of Sargodha, the chowdris were likely to find them through their local contacts. Generally kammis had few contacts in faraway cities or villages that they could rely on to help them find employment and accommodation, but even if they did have friends or relatives in cities like Rawalpindi, Faisalabad, Jhang, or Lahore, those relatives that stayed behind in the village were likely to know where those contacts lived and to tell the chowdri if put under pressure. It was also widely believed that the chowdris could enlist the support of the police to retrieve and even jail-escaped bonded labourers (although this hadn't happened to any of the bonded labourers that I knew). In fact, findings by Human Rights Watch (1995: 70) confirm that the Pakistani police frequently arrest bonded labourers who resist their employers or try to file charges against them. The labourers are often arrested on fabricated charges of theft and the police routinely fail to file complaints against employers (ibid.).

In most cases, however, the chowdris could rely on the immediate kin of their attached servants to make them repay their debts and to prevent them from running away. When a chowdri made a loan to a labourer he generally obtained a guarantor (*zaman*) from among the labourer's household's kinsmen. The guarantor was responsible for repaying the loan, in labour or in kind, if the labourer failed to meet his obligations through abandoning his duties, illness or death. In cases where the house of an attached servant was located on the land of the employer it was generally taken for granted that other able-bodied household members would be held responsible in case the attached servant defaulted on his or her debt-servicing obligations. It was therefore in the interest of kinsmen to prevent their attached relatives from running away, and by extension it was also in their interest to help the attached relatives to service their debts so as to prevent them from wearying of their work and running away, leaving the kinsmen with the burden of servicing the debt in their place. This help took the form of supporting them economically so that they didn't need to take their monthly wages from the chowdri. Thus an indebted attached servant who was paid Rs.10,000 per year could eliminate that amount from his debt if he refrained from claiming his wages and relied on his kinsmen to cover for his basic needs.

The case of Muhammad Hussain Mussalli and his younger brothers serves to illustrate how siblings could cooperate in order to repay a loan from a chowdri. Muhammad Hussain's family was well-off in comparison to most other Mussallis. The family lived in one of the colony schemes established in Zulfikar Ali Bhutto's time and possessed legal title to the plot of land where their three bedroom brick and mud house was situated. Muhammad Hussain's father had made a decent living by dealing in timber and by working as the cleric (*maulvi*) at the small mosque in the centre of Sikander Colony. He had five sons of whom Muhammad Hussain was the eldest, and he had been able to pay for Muhammad Hussain's education through primary and secondary school. Subsequently Muhammad Hussain had spent two years in a nearby Barelvi madrassah where he had become a Hafiz-e Qur'an. After this he had taken up a job as a teacher in a small private primary school owned by a retired air force major in nearby Gullhapur. He earned only about Rs 1,500 per month but between his wages and those of his two siblings of working age, along with some money that his now old and half-retired father continued to receive on Eid, they lived relatively well. The two brothers immediately junior to him had dropped out of school at a young age and worked as daily wage labourers, taking up employment according to the agricultural cycle. Muhammad Hussain's two youngest brothers were still attending school.

Given his educational qualifications, Muhammad Hussain hoped that he might obtain a low-ranking government job or even a job in the military. However he was painfully aware of the fact that he needed the patronage of an influential person for this to happen. The retired air force officer who owned the school where he worked, and who was a relatively influential chowdri in his area, offered to help him out on the military side but was not able to guarantee anything. As for the possibility of a government job Muhammad Hussain complained that despite vague promises none of the Gondal chowdris in the area were helping him. Like many others in similar situations, Muhammad Hussain asserted that the reason for this was that the Gondals didn't want to see locals becoming economically independent because it meant that they would no longer have people around to serve them. He also claimed that many of the chowdris resented the fact that he was an educated Mussalli. He claimed to notice a note of mockery when they referred to him with the honorific Hafiz-ji. This suspicion was not groundless; several chowdris who knew that I was Muhammad Hussain's friend told me that he was an uneducated Mussalli and couldn't possibly have any real understanding of Islam.

When Muhammad Hussain reached the age of 20 his father decided to have him marry and also decided that his second son should obtain a loan of Rs 30,000 from a landlord to finance the wedding. The father hoped that by pooling their earnings and selling some of their livestock all three siblings would be able to repay the loan in under a year. Unfortunately things didn't turn out so well. Within three weeks of Muhammad Hussain's wedding his father had a heart attack and was hospitalised. This meant that the family had to borrow a further Rs 30,000 from the same chowdri in order to cover multiple medical expenses. Even though half of the debt had been incurred on behalf of Muhammad Hussain it was the second son, Muhammad Irfan, who became the servant of the landlord from whom the loan had been taken. One reason that it wasn't Muhammad Hussain was that he already had a job with which he could contribute to repaying the loan. Another reason was that his parents still hoped that with his qualifications and some patronage he might be able to get a better-paid job that would help the family cover its debts. Yet another reason was that, although he was hard-working, Muhammad Hussain was unfit to do the sort of physical labour that would have been expected of him working for a chowdri, and was better suited to work of a more intellectual nature. Muhammad Hussain told me himself that he was incapable of doing heavy agricultural labour. He said that he had once tried harvesting wheat and had ended up severely ill for several days as a result.

Muhammad Irfan was offered the low salary of 24 maunds of wheat per year. At the price of Rs 450 per maund this was equivalent to about Rs 10,800 per year. In addition to this he was to be provided with two meals per day and a cup of tea in the afternoons. His main task was to attend to the livestock but, like most other attached farm servants, he also had to perform a wide range of other tasks which included running errands and attending to guests well into the night. He was required to sleep at the chowdris' farmhouse to guard the livestock from roaming thieves and was only allowed to take a day off if he could provide someone to replace him. Muhammad Irfan frequently complained that his master verbally abused him and never gave him a minute's rest.

Although it was Muhammad Irfan who had to put up with the worst hardship his two siblings worked hard to help him eliminate the debt incurred by the family as a whole. Muhammad Hussain and his other younger sibling pooled their incomes in order to sustain the entire family. This included sustaining Muhammad Irfan so that he wouldn't need to claim the 24 maunds of wheat from his chowdri, diminishing the Rs 60,000 debt by Rs 10,000 by the end of the year. They provided Muhammad Irfan with pocket money of around Rs 300 per month in order for him to purchase basic necessities such as soap and cigarettes. In order to repay the debt more quickly Muhammad Hussain took on extra work in the afternoons after he was finished teaching in the school. At one stage Sufi Ahmed Abbas offered him Rs 600 per month plus daily breakfast and dinner to make the call to prayer three times a day and recite the Holy Qur'an on the loudspeaker in the mornings. Muhammad Hussain accepted the offer and did it for a month, but when the time came to claim his wages Sufi Ahmed Abbas falsely accused him of shirking and failing to perform the call to prayer on several occasions. Sufi Ahmed Abbas decided to deduct the cost of the two daily meals from the monthly pay and told Muhammad Hussain that he would only pay him Rs 300. Muhammad Hussain was outraged by the behaviour of the pious chowdri and left without taking any of the money. After this Muhammad Hussain managed to get a temporary job as a security guard at a construction site in the nearby market town. There he earned Rs 1,000 for his evening and night shifts. The job didn't last for more than three months but during that time Muhammad Hussain became completely exhausted by working all night then teaching from nine in the morning until three in the afternoon. The younger brother contributed to the repayment of the debt by working during the tangerine and wheat harvests and by finding work in construction outside the peak periods in the agricultural cycle.

With great effort the family managed to decrease its debt by Rs 10,000 within a year. In the second year Muhammad Irfan shifted to another chowdri who offered to pay him Rs 15,000 per year and purchased the remaining debt of Rs 50,000 from the previous employer. In that year the debt was further reduced by Rs 15,000 through the ongoing combined efforts of the three siblings. Muhammad Hussain continued teaching and taking up supplementary jobs in the afternoons. In the meantime he continued searching for patronage in order to obtain a formal sector job. During this time his younger brother continued doing daily wage labour. The remaining Rs 45,000 was repaid shortly after the end of this second year when the family sold a buffalo and some goats that they had been raising in order to cover the debt. When the debt was finally repaid, Muhammad Irfan returned home and Muhammad Hussain invited me to his house to celebrate the end of their 'punishment' (*azaab*) with a large meal.

After this, Muhammad Irfan, who felt physically and mentally drained after his two years of hard labour, took several months off work to recover. Muhammad Hussain managed to get a government job in southern Punjab with the National Logistics Cell through the patronage of the retired air force major. In the end, however, the job proved to be a great disappointment. The National Logistics Cell was a notoriously corrupt branch of the army supposed to combat smuggling of various sorts on Pakistan's main highways. As it turned out patronage was not only necessary to obtain the job but also in order to keep. While he was on the job Muhammad Hussain witnessed one of his superiors taking a bribe from a heroin smuggler. When the superior in question was later questioned about this by his own superiors, he summoned Muhammad Hussain to lie that he hadn't seen him taking the bribe. But when Muhammad Hussain was asked to swear on the Holy Qu'ran that his superior hadn't taken a bribe he refused to do so. As a result his superior, who actually got away with taking the bribe, transferred Muhammad Hussain to a remote location in Sindh. Muhammad Hussain told me that without a superior to look after his interests it was impossible not to be pushed around in such cases. Furthermore his pay was only Rs 3,000 per month and he spent almost Rs 1,000 of this when he returned to visit his family and wife. In the end he decided that he would be better off earning only Rs 2,000 somewhere nearer to home where he wouldn't be homesick and wouldn't have to compromise his integrity by turning a blind eye to heroin-smuggling and human trafficking.

As a result of all of these experiences Muhammad Hussain grew increasingly bitter about the daily injustices faced by the majority of Pakistanis. He would often talk about the way that the landlords were holding the country back and about how all of their activities violated the basic tenets of Islam. At one stage he even joined a local worker and farmer (*kisan aur mazdoor*) movement headed by a retired army colonel in Sargodha. Muhammad Hussain told me that the movement gained a number of followers from within the district and that rallies and other functions were held on a regular basis. Muhammad Hussain became a member and discretely went around the area of Bek Sagrana getting people to sign up and make small contributions to the movement. He claimed that he managed to arouse the interest of some people but that the majority appeared to think that the whole exercise was futile. The doubters were eventually proven right when the retired colonel ran away with all of the money that he had allegedly been collecting for the movement. Obviously this episode left Muhammad Hussain even more embittered.

Other cases

Instances whereby families cooperated in order to repay a loan from a landlord like the one described above were quite common. For example, one labourer who belonged to a family of impoverished non-Gondal zamindar tenants also managed to pay off a loan from a chowdri, thanks to the income and labour of three able-bodied siblings. The labourer had obtained a loan of Rs 50,000 from a chowdri in order to cover his father's medical expenses. Thanks to the labour of his siblings and the sale of some livestock the loan was repaid after two-and-a-half years of toil.

There were instances, however, when the joint-servicing of debts didn't go so smoothly and could result in bitterness and conflict within households. In one Mirasi household the older of two brothers secured a loan of Rs 15,000 from a Gondal chowdri in order to cover joint-wedding expenses for himself and his younger brother. The brothers agreed that they would service the debt by working for two different landlords. In the end, however, it was the older brother who ended up having to pay the greater part of the loan because his younger brother left his master after only six months of work and then spent several months wasting time without permanent employment. Thus the elder brother worked for a year-and-a-half for a chowdri and repaid Rs 11,000 while his younger brother repaid only Rs 4,000. This situation created a great deal of bitterness between the two brothers but they still continued to live in a joint household with their parents.

When indebted servants had no able-bodied siblings or parents to assist them, repaying a loan could take an entire lifetime, and rather than see their debts diminish, they were likely to see them grow. Repayment was even more difficult if a household was afflicted by additional crises such as illness or drug addiction. In one instance, a Mussalli who was a heroin-addict had given his seven-year-old-son Fazi over as a servant to a Gondal chowdri in exchange for Rs 6,000 per year and various small cash advances. The family had come to depend economically on the young boy because of the father's addiction and the absence of other able-bodied males in the household. By the time Fazi turned 17, the debt had risen to Rs 20,000 as the household was continually short of money. The father had sold much of the furniture in the house in order to finance his heroin addiction. The debt further increased by Rs 30,000 when Fazi and his sister got married. At the time of fieldwork, the outstanding amount totalled Rs 50,000 and the Ranjha chowdri had decided to cut off any further flow of credit because it was becoming unlikely that the loan would ever be repaid.

The situation for Fazi had, however, become more hopeful as he now had a younger sibling of working age who was working in a vermicelli (*sevian*) factory in Sargodha and was contributing to family finances. The father had also ceased to be a burden since going to Islamabad where he found work in a market carrying people's shopping bags to finance his heroin addiction himself. In addition, a charitable and pious Ranjha chowdri had given Fazi a baby buffalo to rear, such that when the animal was fully grown they would share the proceeds equally. An adult buffalo could be sold for about Rs 80,000, allowing Fazi to reduce his debt by as much as Rs 40,000 from his share. However, there was an unfortunate setback: his sister contracted tuberculosis and her in-laws sent her back home because she had become a financial burden. Medical treatment was expensive and Fazi, who was desperate to repay his debt, complained that it was the duty of his sister's in-laws to take care of her. He threatened not to make any financial contributions towards her treatment.

Conclusion

This chapter has illustrated how despite the proletarianisation of the workforce, landlords were still able to enforce exploitation what has been termed 'neo-bondage'. To conclude I wish to emphasise the fact that it was by making the repayment of loans the collective responsibility of households that the Gondals could recover the debts of their labourers. Some bonded-labourers told me that it was because their relatives would be made to pay off their debts if they ran away that they didn't do so. Collective responsibility also meant that the entire household of an indebted labourer had an interest in the latter working to repay his debts because if he didn't, the burden of doing so would fall on one of them. As a result they helped their relatives service their debts by pooling their labour, meaning that labourers with several able-bodied siblings – and parents – could repay large loans within three to four years. Another consequence was that the kin of indebted labourers had a direct role in enforcing the repayment of debts. It was often they who made sure that bonded labourers attended their duties and that they didn't run away. However, as the case of the two Mirasi siblings discussed above illustrates, they weren't always successful in doing so. This was particularly the case when there were quarrels within the kinship group. Thus when Fazi – mentioned before – once had a dispute with his paternal uncle, he ran away to Sargodha and left his uncles to face the consequences. Since he had no siblings, it was this uncle – who worked as tractor driver for the same landlord – who was accountable for him. In order not to have to personally service his nephew's debt it was he who found out where Fazi was and who brought him back to his master.

A broader consequence in terms of labour assertiveness of this way of enforcing debt-servicing was that class conflict was 'transferred into the domain of "innate"/"immutable" and historically longstanding identities (actual/fictive kinship, caste) supported by religious concepts of the sacred' (Brass 1999: 28). Thus, instead of struggling with the landlords for better working conditions, kinsmen – such as Fazi and his uncle or the two Mirasi brothers mentioned before – ended up struggling with each other. This compounded the fact that labourers as a whole were already divided by caste membership and by the logic of patronage which encouraged practices such as snitching. Caste divisions meant that labourers showed little solidarity towards each other. Labourers showed particularly little solidarity towards indebted Mussallis whose fate they attributed to their prodigality and lack of self-control. Several kammis told me that it was because Mussallis were spendthrifts who spent too much on their weddings, drank alcohol and bought flashy clothes that they became indebted. Moreover as indicated in chapter one, people's need to ingratiate themselves with the Gondals in order to get access to patronage and protection meant that they often resorted to snitching on each other. When I asked people what would happen if the relatives of an absconding bonded labourer were unwilling to help find and retrieve him, I was told that there would be no shortage of sycophants (*chamchas*) ready to assist the landlords and reveal his whereabouts.

The fact that the Gondals controlled access to housing was the key factor that allowed them to exert pressure on indebted labourers through their kin. Had they not been dependent on landlords for housing, but also access to fodder and firewood, kammis are less likely to have taken so much trouble ensuring that their indebted relatives serviced their debts. Ultimately it was the fact that the Gondals controlled the state which effectively deprived kammis of secure housing rights. It was this that made it impossible for them to seek protection from the courts and from the police when landlords sent gunmen – or even the police – to forcefully evict them. But it was also landlord control over the state that often caused kammis to become indebted in the first place. As illustrated in chapter three, it was because the Gondals had effectively plundered the state that public hospitals didn't work and that, as a result, kammis needed to get loans in order to obtain medical treatment.

3

ELECTORAL POLITICS AND THE REPRODUCTION OF INEQUALITY

Reproducing dominance

This chapter provides an account of how the Gondals of Bek Sagrana were able to consolidate their political and economic clout and transform their local dominance in terms of numbers, landownership and education into 'decisive dominance' (Srinivas 1987) starting in the 1960s. Prior to this a single Makhdoom household in the village of Nawabpur had controlled local, national and provincial assembly seats while the influence of the Gondals had been largely restricted to their villages. However the green revolution and political shifts under General Ayub Khan and under Zulfikar Ali Bhutto gave wealthy cultivators from numerous castes – such as the Gondals – an electoral advantage over feudal landlords whose power rested solely on hierarchical patron–client ties and on the types of debt-relation described in the previous chapter. In the case of Bek Sagrana, as in many other cases, this advantage over feudal landlords was enhanced by Bhutto's mobilisation of the masses against them through the promise of land and homestead reforms.

I suggest that the growing political clout of the Gondals from the 1960s onwards was representative of the trend identified by Shahid Javid Burki (1988) according to whom wealthy, 'middle-ranking' farmers owning between 50 and 150 acres of land gained political power, often at the expense of the landed aristocracy. Although it is true that the latter continued to play a prominent role in Pakistani politics, particularly in the Southern Punjab and Sindh, the mobilisation of larger sections of the peasantry did signal a shift away from feudal politics. Asad Sayeed (1996: 272) argues that in 'demanding resources and rights from the state the middle and small peasantry employs the means of populist politics rather than a one-to-one patron–client approach that the larger landlord employs'. In other words the middle and wealthy peasantry had to join forces rather than act individually like the feudal landlords who could rely almost solely on their personal dependants for political support.

This chapter examines what the rise of a more populist form of politics meant for the reproduction of local patterns of power and authority and, more generally, what this meant for democracy in Pakistan. I argue that by pursuing the particularistic interests of their kinship groups, clans and factions – with the help of higher-level politicians backed by the military establishment – 'populist' landlord politicians reproduced concentrations of economic and political power and even created new ones. As such, they undermined the state's role as a provider of public goods such as health care, justice, housing, and even education. Thus, contrary to Partha Chatterjee who believes that popular freedoms advance through the ad hoc clientelistic practices of what he calls 'political society', I argue that these practices are in fact central to the reproduction and consolidation of existing forms of domination. Moreover I suggest that elites often benefit less from the rule of law – which Chatterjee likes to term 'Bourgeois' – than from its absence and that its absence is particularly stark in Pakistan where military interventions have repeatedly undermined it.

I question whether the failure by Pakistani politicians to engage in what Wilkinson (2011) calls 'programmatic politics' – in which politicians act with a view to enhancing the universal public interest rather than particularistic interests – is related to the fact that, in Migdal's (1988) terms, Pakistan has a strong society and a weak state. In different ways both Anatol Lieven (2011) and Matthew Nelson (2011) make this point when they argue that the importance of clan-based patron–client ties in Pakistan mean that leaders redirect public resources to benefit kin, friends and clients thereby subverting central government's attempts to implement both policies and legislation.

In this chapter I suggest that the perpetuation of parochial politics in Pakistan is related to the alliance between the landed classes and the military rather than to the persistence of 'tribal' power-structures. I argue that by preventing free and regular elections and by putting a break on attempts at land reform, the military helped entrench the landed class and rural patronage structures in Pakistani politics. As Javid Hassan (2011) has shown, local clientelistic power structures have remained in place because authoritarian postcolonial regimes in Pakistan have used them to buttress their rule, much like the British colonial state previously did. Moreover the military helped keep politics parochial by undermining the impartiality of state institutions – including the judiciary – and by effectively preventing politicians from acting as legislators and policymakers.

Authoritarianism and parochial politics

The persistence in Pakistani politics of traditional power structures organised around the parochial loyalties of kinship, clan and faction cannot be explained without reference to the continuation of colonial traditions of indirect rule through landed notables. Although the colonial and post-colonial states clearly didn't create and sustain these traditions out of nothing, they played a crucial role in perpetuating them and in preventing the emergence of more inclusive forms of political mobilisation. Even now that Pakistan is on the road to establishing procedural democracy, it can at best only be described as what Chandra (2007) calls a 'mediated democracy' in which only a minority of voters are autonomous and control the votes of the rest.

Imperial rule in the Punjab was premised upon the idea that political stability was best ensured by securing the support of rural notables and, more broadly, of the 'agricultural tribes'.[1] To this end the British went about co-opting rural elites by establishing a 'loyal class of "hereditary landed gentry" by awarding pensions, titles, and land grants in the developing canal colonies, and by confirming ownership of large semi-feudal estates in Western Punjab' (Wilder 1999: 69). Imran Ali's (1988) study illustrates how the canal colonies were created to facilitate accumulation by the colonial state and to reinforce the political and economic position of the landed class. Focussing on the strategic concerns of the colonial government, Yong (2005) and Saif (2010) show how it sought to retain the loyalty of the Punjabi landed class because it depended on it for military recruits, particularly after the Mutiny.

In addition to being given land grants, selected notables from within the rural elites were also entrusted with key aspects of administration both informally and formally. Wealthy peasant proprietors were given revenue collecting functions as *zaildars* and *lambardars*: the zaildar was at the head of an agricultural tribe and in charge of collecting land revenue within a *zail* – a jurisdiction that was roughly congruent with the settlement pattern of a single agricultural 'tribe' – and the lambardar was a village headman

---

[1] The bias towards rural notables by imperial authorities is clearly illustrated by the passing of the Punjab Land Alienation Act of 1900, which prevented the transfer of land from indebted members of 'agricultural tribes' to money-lending members of the 'trading classes'. Because of the importance of land in determining social and political status, imperial authorities wished to ensure that it would not be transferred to the 'trading classes', whose loyalty to British rule was in doubt. See Van den Dungen (1972).

entrusted with collecting revenue for a particular village. These positions were hereditary and entitled the persons who held them to five percent of the revenue that they collected as well as significant amounts of state patronage.

The power of agriculturalist tribes was further entrenched by the Panchayat Act of 1912, in which informal village councils (*panchayats*) were given formal recognition as dispute resolution forums, giving their decisions legal sanction. The colonial administration claimed that these tribunals provided cheaper and faster forums for obtaining justice in minor civil and criminal cases, but the intention was to further entrench the landlords within the Punjab colonial state. To this end the authorities ensured that the Panchayats would continue being dominated by landlords by making membership in them contingent upon owning property. Because of this dominance, the judicial decisions made by these bodies tended to protect the interests of the locally dominant landed clan (see Chaudhary 1999) at the expense of outsiders and of the landless. Moreover the funds granted to panchayats for minor development projects – including the construction of roads, paths and buildings – were expressly meant to benefit landlords.

Additionally agriculturalists were granted positions within formal bureaucratic institutions and by 1917 roughly half of the province's *patwaris, naib-tehsildars* and policemen were agriculturalists (Hassan 2012: 127). In March 1917 this led the Chair of the Punjab Legislative Council to approvingly observe that 'there is no province in India which can compare with the Punjab in the extent to which zamindars are employed in government service from the rank of patwari and constable upwards' (quoted in Hassan 2012: 129). Finally, agriculturalists were allowed to capture elected positions within district councils and provincial assemblies thanks to property qualifications on the right to stand for elections and to vote. The colonial government held that showing preferential treatment to agriculturalists was justifiable because of their services during the First World War, their contribution to the revenues of the province and their contribution to the stability of the province.

The position of these rural notables as intermediaries with the Imperial authorities entrenched them within the administrative and political structure of the Punjab both at the expense of urban constituents as well at the expense of their subordinate clients, who became dependent upon them for all forms of mediation involving state institutions. Their influence within the colonial state allowed them to collectively pursue their interests as a class and to engage in rent-seeking activities.

Under late imperial rule the urban constituencies in the Punjab were composed predominantly of Hindu traders and a Muslim middle class that was becoming increasingly involved with the expanding colonial government as lawyers, physicians, engineers, and teachers. As urban Muslims became increasingly involved in government, their expectations for greater political influence started to grow, leading many of them to join the All-India Muslim League founded in 1906. However, in the politically dominant rural Punjab, where the majority of the population still lived, the Muslim League largely failed to establish a popular base of support. Rural notables joined forces on a non-communal basis under the banner of the Punjab Unionist Party, largely in order to protect their landed interests which they perceived to be threatened by the political agendas of both the Muslim League and the Congress Party (see Jalal 1995). As a result, the overall support for the Muslim League in the Punjab was insignificant until 1946, when Jinnah obtained the support of opportunistic rural notables who perceived that independence was imminent, and that it was therefore in their best interests to join the League.[2]

The fact that the rural notables of the Punjab took over the local leadership of the Muslim League on the eve of independence meant that the party was unable to establish a rival structure of power that bypassed the rural elite. This, combined with the fact that landlords had significant power over the bureaucracy, meant that they remained in their position as local intermediaries, and the Muslim League never established itself as a disciplined political party with a direct base of popular support.[3] Moreover, the fact that the national leadership of the Muslim League had no base of support in the national territory (because it hailed predominantly from within the territory that became independent India) further hampered the party's efforts to establish itself. The Muslim League leadership's lack of popular support meant that it became increasingly reliant upon the civil–military bureaucracy to retain its hold on power, and it succeeded "in maintaining the facade of parliamentary democracy without holding national elections from 1947 to 1958" (Wilder 1999: 18). This facade of parliamentary democracy gave way to martial law in October 1958 when the Muslim League, backed by the West Pakistan dominated civil and military bureaucracies, imposed central rule on East Pakistan, suspending all political activity there. The Muslim League and its allies within the civil

---

[2] Ayesha Jalal notes that desertions from the Unionist camp 'had little to do with the Punjab League and its invisible organisation, and more to do with men anxious to win tickets . . . "Sensational stories of conversion to the League" took the provincial League leaders by surprise (Jalal 1995: 145).

[3] See Syed (1989) for an account of how factionalism amongst powerful landlords in the Muslim League created a great deal of political instability between 1947 and 1955.

and military bureaucracies thwarted democratic forces through the imposition of Martial Law by President Iskander Mirza, which shortly led to a takeover by Army commander General Ayub Khan.

In a pattern that was to be replicated by future military rulers in Pakistan, Ayub Khan banned political parties, passed legislation that was selectively used against political opponents, and erected a local government scheme that was to serve as a democratic facade for his rule and an easily controlled electoral college for the presidency. He also modified the constitution in order to replace Pakistan's parliamentary system with a strong presidential one that severely curtailed the powers of elected assembly members. These policies – and similar ones enacted by his successors – undermined the role of politicians as legislators and turned them into mere brokers for state resources at the local level. This arguably exacerbated their tendency to focus on parochial issues rather than on issues of national interest. Moreover by banning political parties – and thus institutional funding for political candidates – Ayub Khan ensured that only candidates with private money and influence would be able to contest elections; this arguably further entrenched landlords into the political system.

Furthermore, Ayub Khan tried to ensure strong economic growth and rapid industrialisation by adopting neo-liberal economic ideologies according to which the key to rapid industrialisation was cheap rural migrant labour. This meant that restrictions already in place against labour unions were strengthened. According to this model of economic growth significant land reforms would keep the rural population on the land and therefore restrict the flow of migrant labour into the cities (see Candland 2007: 48). Thus, the ceiling that General Ayub Khan imposed upon landownership was aimed at breaking the power of the minority of extremely powerful aristocratic (*ashrafi*) landlords rather than towards a significant redistribution of the land. It was also designed to encourage commercial agriculture by forcing large landholders to increase their productivity as a result of reduced landholdings. The political clout of the large ashrafi landlords, and their ownership of land, were disproportionate to their total numbers, and General Ayub Khan sought to curb their influence in order to enhance state power (see Nasr 2001). To this end he sought to create a rural constituency from amidst the ranks of the middle and gentry landholders who constituted the majority of landowners in the Punjab. According to one study, 62 per cent of West Pakistan's Basic Democrats belonged to this class of landlords (Inayetullah 1964: 51–61). Therefore while basic democrats didn't have as much wealth as the aristocratic landlords that the Ayub regime sought to displace, they nevertheless still represented similar class interests.

It is from among these ranks that the Gondals of the district of Sargodha emerged to political prominence. Under Ayub Khan the representation of this class of landholders in the national assembly 'went from 9.5 per cent in 1955, up to 44.7 per cent in 1962, and then back down to 29.8 per cent in 1965' (Wilder 1999: 74). Both the entry of this class into formal politics under Ayub's regime as well as the substantial economic gains that they made as a result of Pakistan's first green revolution during this period turned them into a significant force in Punjabi politics.

Over the decade during which General Ayub Khan was in power, Pakistan witnessed rapid industrialisation and economic growth. The rewards of this growth, however, were highly unevenly distributed. According to one estimate from 1968, 'industrial ownership was concentrated in twenty-two families' (Jones 2003: 145). For the masses that had moved to the cities as a result of the rapid mechanisation of agriculture during Pakistan's green revolution, and from the eviction of tenants that this entailed, the rewards of growth were far less tangible. Moreover, General Ayub Khan's decade in power had failed to provide sufficient resources for the country's human development. As a result, when the Ayub era came to a close only 15 per cent of the population could be classified as literate. The country also had one of the lowest ratios of doctors per rural inhabitant in the world with one doctor for every 24,200 people (Burki 1980: 47). The failure of such growth to benefit the masses, together with the severe restrictions on political freedoms, eventually resulted in widespread agitations. In East Pakistan it resulted in the rise of Mujibur Rehman's Awami League demanding provincial autonomy and democracy, while in West Pakistan it resulted in the rise of Bhutto's Pakistan People's Party (PPP) advocating a socialist revolution with a programme for the nationalisation of industry and major land reforms.

The principal support for Bhutto initially came from urban left groups such as organised labour, and students but also from middle class professionals and the intelligentsia. The promise of land reforms together with the promise of homestead reforms[4] and the abolition of corvée labour (*begaar*) also enlisted the support of tenants, landless labourers and village artisans (kammis). Furthermore, the poor rural voter hoped that the PPP would pursue its interests against 'the classes and institutions that had weighed upon him so heavily and for so long – the zamindariat, the police, the bureaucracy and the courts' (Jones 2003: 362). As Jones (ibid.) has convincingly shown, the 1970 elections – the first national elections based on universal adult franchise, which saw Zulfikar Ali Bhutto rise to power – were the first time in the country's history where the have-nots erupted onto the political

---

[4] Homestead reforms were aimed at ensuring that house tenants could no longer be arbitrarily evicted from their homes in the villages that belonged to landlords.

scene. According to Jones these elections not only saw the emergence of horizontal, class-based ties of allegiance among urban voters but also among rural voters, where historically the masses had 'been politically quiescent, conditioned by tradition and the local structure of power to follow their customary leaders – ashrafi-gentry zamindars, pirs, and clan biraderi heads' (Jones 2003: 356). Because of the importance given to land and homestead reform in the PPP's electoral campaign, party support patterns during the elections were often divided horizontally along class lines between those who owned land and those who did not.

In one case study, Jones examines voting patterns in a village in the Punjabi district of Gujranwala that was divided into two factions of Tarar Jats whose leaders had chosen to support opposing candidates from two different parties that were running against the PPP. He shows that while one of the factions successfully delivered votes to their candidate, the other faction split into two with tenants, kammis and middling zamindars opting out of the faction and giving their vote to the PPP. Jones shows that even though many of the tenants in the village belonged to the dominant Tarar Jat biraderi their voting behaviour was determined by the prospect of political and economic empowerment at the expense of ascribed biraderi-based loyalty. Moreover, landless labourers and kammis of this faction also broke off their traditional ties of allegiance with their factional leaders. Although one of the factions remained intact, the fact that one of them became horizontally divided was itself a significant development that was repeated throughout the Punjab. In other instances Jones reports that the PPP encouraged significant numbers of rural tenants 'to form their own organisation to support the PPP, despite threats and economic pressure from landed interests' (ibid.: 365). Thus, for example, Jones recounts that in one instance in a village 15 km southwest of Lahore tenants under the leadership of a man called Sandhu Jat secretly and by consensus decided to vote for the PPP. Furthermore, although several elite landlords – including a branch of the influential Noon family in Sargodha – voted for the PPP, several others forgot their differences and allied in a horizontal elite pattern against the threat presented to their landed interests by the PPP. Jones relates another instance, once again in the district of Gujranwala, where the traditionally rival Chathas and Awans united against the threat of the PPP.[5] Such patterns apparently became very common during the National Assembly elections 'when parochial elite candidates made desperate (and

---

[5] This was a significant departure in Gujranwala, where *Awans* and *Chathas* had consistently supported opposing candidates as a result of a long-standing enmity.

occasionally successful) horizontal alliances in order to defeat the unexpectedly strong pro-PPP vote produced in the NA election' (ibid.: 366).

Despite a start that promised political empowerment for the masses, Bhutto's regime soon revealed itself to be just as authoritarian as its predecessors. Many of his policies, including his reform of the civil services and his nationalisation of large-and middle-sized industries, served to extend his personal powers and his capacity for patronage. Rather than strengthen the PPP into a coherent party organisation with established internal rules and procedures Bhutto increasingly concentrated power in his own hands. Moreover Bhutto, like all other political leaders in Pakistan, found that he needed to make alliances with powerful members of the landlord class. Despite all of this, the regime did manage to at least temporarily further the interests of the poorer segments of rural society through tenancy, land and homestead reforms. For example, during his time in power, reports suggest that landlords found that the judiciary had become very receptive to suits brought by tenants seeking redress for eviction. Thus, Herring reports that in 70 per cent of all cases of eviction, judgements ordered the restoration of tenants (Herring 1983: 117). With respect to land reforms it appears that landlords found loopholes in the new legislation, which set the land ceiling for irrigated land at 150 acres and that for rain-fed land (*barani*) at 300 acres. Chief among these loopholes was the fact that the ceilings were set for individual holdings rather than family holdings, allowing many landlords simply to transfer their lands to close family members. As a result the impact of the land reforms remains unclear to this day, although it appears that some of the largest estates were affected. For example, I was told by a prominent aristocratic landlord in Sargodha that he had lost a significant amount of land to tenants but that despite this he still controlled more than 1,000 acres. Similarly the Makhdoom lineage of pirs, who were once the most prominent ashrafi family in the area where I conducted my fieldwork, also lost land to their tenants and several kammi families, who now possessed legal title to it. Nevertheless, the Makhdooms still possessed land far above the ceiling imposed under Bhutto's reforms.[6]

Another set of reforms implemented during the Bhutto era whose impact is contested (see Rouse 1983) was that of homestead reform. In rural areas villagers usually lived in houses and on land that belonged to a local landlord. In exchange for a place to live, house tenants were expected to provide corvée labour (begaar) and political support to their landlord. Bhutto's

---

[6] On the other hand, the Gondal landlords whom I studied, and who possessed several hundred acres of land, had not been similarly affected by some loss of land under the reforms.

reforms aimed to free house tenants of such ties of dependence by granting them legal titles of ownership over their houses. Another related programme was one where colonies were created on state-owned land, and five *marla*[7] plots of land were granted to people who had previously lived on land that belonged to a landlord.

With respect to the first programme Rouse has noted that 'legal titles do not seem to guarantee ownership by the resident' (ibid.: 323), which she goes on to illustrate with a case from 1979 in which a domestic employee in the village of Sahiwal who had quarrelled with her mistress was summarily ordered to vacate her house. More than 25 years later most kammis in Bek Sagrana believed that the Gondal chowdris could still arbitrarily evict them from the houses they occupied in the village. The inhabitants of housing colonies were more secure in their ownership over their housing plots and therefore had a greater degree of independence from the landlords. During local body elections in 2005 I was told by chowdri politicians that colony residents didn't simply follow their dictates as to whom they should vote for, but had to be wooed through patronage and cash. Moreover, colony residents were freed from having to carry out begaar for any landlord. Nevertheless, in 2005 in a colony named Sikander Colony after the Gondal chowdri who owned most of the land adjacent to it, people still gave their vote to the faction supported by Chowdri Sikander for fear that he might mobilise strongmen to evict them if they didn't.

Despite the highly imperfect legacy of Bhutto, most landless villagers in the Punjab and elsewhere still considered the PPP to be 'their' party, and those who lived through Bhutto's time in power recalled it as an era of hope and empowerment. This is true despite the fact that nowadays the PPP is largely known to be the party of large landlords who control blocks of voters through their local power and landholdings, and who sometimes still deploy its populist rhetoric to swing votes their preferred ways. Despite this, both kammis and former tenants told me that if they could freely choose they would vote for the PPP (which at the time was still headed by Benazir Bhutto) because of what Zulfikar Ali Bhutto had done for the poor.

Although land, tenancy and homestead reforms were never perfectly implemented under Bhutto's regime, much of what had been achieved was reversed by General Zia-ul Haq. The need for a constituency meant that Zia-ul Haq, like his British and Pakistani predecessors, reached out to the rural notables whose property he pledged to protect from a second round of land reforms that had been planned by the Bhutto government. Landlords

---

[7]A *marla* is a traditional unit of land measurement in South Asia. One marla is equal to 25.3 square metres.

were given the green light to add to their existing holdings and even to recover some of the land lost under Bhutto. Moreover, the judiciary ceased to be sympathetic to the claims and issues of tenants and their cases were ignored by the courts, who once again sided with the rural elites. Ever since Bhutto the issue of land reform has never resurfaced and Musharraf's government actively encouraged the consolidation of large landholdings by incentivising the corporatisation of agriculture.

On the political front General Zia-ul Haq's regime ensured that the sort of mass mobilisation which had occurred in Bhutto's time could not take place. He did this through the large scale repression of the political opposition spearheaded by the PPP and by holding non-party local body elections. Like his predecessor General Ayub Khan and like his successor General Pervez Musharraf, Zia encouraged candidate-based as opposed to party-based electoral campaigns. According to Muhammad Waseem,

> Zia attempted to revert to the early colonial mode of district politics in which local influentials got elected into the legislatures on the strength of their respective support bases in the locality characterized by the ties of tribe, caste, faction or tenurial relations. The obvious target of this policy was the PPP which had continued to enjoy mass popularity. . . . The 1985 non-party elections localized politics, reinvigorated biradri[sic.], cut across the potential – though non-active – lines of party support and decisively shifted political initiative towards electoral candidates.
> (Waseem 1994: 15)

In this manner the parochial political issues and factions that Zulfikar Ali Bhutto had temporarily succeeded in overcoming returned to dominate Pakistani politics.

Zia-ul Haq's regime did not simply return things to the status quo however, and various new features drastically altered the nature of politics and resulted in unprecedented levels of violence. Among these new features were heroin and cheap weapons. General Zia-ul Haq played a major role in the cold war by supporting the Afghan Mujahideen's resistance against the Soviet invasion of Afghanistan, and Pakistan received substantial amounts of aid from the United States for this participation. As a result of this policy 3 million Afghan refugees entered Pakistan (Jalal 1995: 108) and the Federally Administered Tribal Areas (FATA) emerged as a major hub for the production of weapons to supply the deadly commerce of the Afghan war. Weapons including cheap Kalashnikovs became widely available throughout the country. In 2005, a Kalashnikov could easily be purchased on the open market for as little as Rs 3,000. Moreover, a parallel drugs economy based predominantly upon the traffic of heroin, widely believed to be linked with the army's notorious Inter-Services Intelligence (ISI) wing, emerged. Thus while heroin addiction was negligible before 1979, estimates suggest that by the year 2000 there were between 1.5 million and 4 million heroin addicts in Pakistan.[8] As will be shown later, both the arms and the drug trades made for increasing levels of violence and the widespread seepage of crime into politics when politicians, in collusion with the police, became involved in these lucrative trades not only in the cities but also throughout areas of the rural Punjab. Furthermore, even though control over local toughs (*goondas*) was already an integral part of politics prior to Zia's time, as Hamza Alavi (1971) has shown, evidence seems to suggest that it came to play an even more central role during and after the Zia years. During fieldwork, politicians openly told me that in order to control their locality and to win elections it was imperative for them to have strong ties with local goondas. These goondas constituted heavily armed groups who were frequently involved in various forms of trafficking and who could be called upon to intimidate opponents and voters as well as to occupy land illegally. As Brass (1997) demonstrates for the Indian state of Uttar Pradesh, political power in Pakistan was increasingly based on force rather than patronage. This issue will be explored throughout the book but will be given particular attention in chapter four.[9]

---

[8] See *New York Times* article by Bearak (2000) who reports the difficulty of obtaining accurate figures on the number of heroin addicts due to the low priority given to social science research in Pakistan.

[9] General Zia's period in power is also widely recalled for the 'Islamisation' of Pakistan's laws and the implementation of the Hudood Ordinances, which have been widely condemned by human rights groups for causing increased violence towards women, minorities and other powerless groups. Under these laws the relatives of a murdered person were entitled to settle their case with the offending party through the payment of blood money (*diyat*). This resulted in an increase in 'honour killings' whereby wives, daughters or sisters who were suspected of tarnishing the honour (*izzat/ghairat*) of their families were put to death by relatives. Under these laws a woman's legal guardian, usually her father or husband, was entitled to forgive that woman's murderer. What often happened as a result was that husbands who suspected a slight to their honour could get a relative to kill their wife and then forgive them. Moreover, although less has been written on this issue, the possibility of settling murder cases through the payment of blood money meant that powerless individuals could be pressured into forgiving a murder with minimal compensation. Thus, for example, a landlord might kill a tenant or a servant and then pressure his family through threats of eviction or even violence into agreeing to a settlement.

General Zia's rule came to an abrupt end on 17 August 1988 shortly before the non-party elections scheduled for November of that same year, when he died in a mysterious plane crash along with several of Pakistan's top generals and the United States Ambassador. The Supreme Court subsequently declared that General Zia's stipulation that elections should be held on a non-party basis was unconstitutional, and elections were held with the full participation of political parties. The 1988 election was essentially a contest between the supporters of General Zia, who united in a coalition under the banner of the Islami Jamhoori Ittehad (IJI) backed by the military establishment, and the PPP. After Zia's death, continued military rule was unlikely to meet both domestic and international acceptance, so the military formally transferred power to civilians while retaining control over foreign and defence policy. Subsequently between 1988 and 1996 it used a combination of bribery, coercion and electoral manipulation in order to secure its interests.

Although the PPP under the leadership of Benazir Bhutto won the elections with 38.5 per cent of the national vote, rigging deprived it of the majority necessary to form a government. Army Chief General Mirza Aslam Beg finally allowed her to form a government on condition that she ceded foreign, security and economic policy to the military.

In the Punjab the IJI obtained much of its support from the conservative urban middle classes, urban traders and industrialists, who had been alienated by the PPP and had greatly benefitted from state patronage and from economic liberalisation under General Zia. Following the death of General Zia the man at the head of this constituency was Mian Nawaz Sharif, who had held various influential positions in the Punjab government including that of Finance Minister from 1981 to 1985 and Chief Minister from 1985 through 1990, when he became Pakistan's Prime Minister. Nawaz Sharif's rise to power was significant because he represented the growing political clout of the Punjabi urban middle classes to which he personally belonged. His time as Finance Minister and Chief Minister had allowed him to develop strong ties with the provincial bureaucracy, which he had succeeded in making subservient. He achieved this by filling thousands of government jobs with his supporters and by transferring officers who showed excessive independence. As a result, Nawaz Sharif put himself in a position to dole out patronage on a 'scale never before witnessed in Punjab politics' (Wilder 1999: 138), thereby firmly entrenching his power at the expense of the PPP. Consequently, when Benazir Bhutto became Prime Minister in 1988 her government was essentially powerless in the country's most powerful province. This situation resulted in a great deal of political acrimony and instability, the eventual premature dismissal of Benazir Bhutto's government 20 months later and in the eventual rise to power of Nawaz Sharif in 1990.

Over the next nine years leading to the military takeover by General Pervez Musharraf in 1999, power went back and forth between the PPP and the Pakistan Muslim League Nawaz (PML-N) without any government ever completing its term in office and with no government lasting for more than three years. Both parties repeatedly succumbed to the military's divide and rule tactics. Their governments were repeatedly dismissed by presidents handpicked by the military. After Benazir Bhutto's fall in 1990, Nawaz Sharif held the reins of power for only three years before being himself unseated by President Ghulam Ishaq Khan on charges of mismanagement and corruption. In 1993 Benazir Bhutto returned to power, only to be unseated again in 1996 by President Farooq Leghari, once again on charges of corruption and mismanagement. In February 1997 Nawaz Sharif returned to power and was superseded two years later by General Pervez Musharraf who broke the pattern of his civilian counterparts and held on to power for approximately nine years.

Wilder has attributed the political instability of this decade of civilian rule to patronage politics, arguing that the 'problem with patronage politics is that there is never enough patronage to keep the majority of voters happy' (ibid.: 235). In an authoritarian system where winners can take all – because state institutions, including the judiciary, can be bent to fit the purposes of the executive – political opponents are denied access to patronage and are often victimised. As a result, they resort to agitation and other techniques in order to overthrow the incumbent government.

Following this decade of unstable civilian rule, General Pervez Musharraf imposed military rule once again in October 1999.[10] General Musharraf's policies and his stance towards politicians and political parties were similar to those of his military predecessors, Generals Ayub Khan and Zia-ul Haq. He believed that politicians were the cause of Pakistan's woes because they dedicated themselves to rent-seeking and to their private vendettas (see chapter 4) instead of to governing. Like his predecessors he attacked political parties and implemented a devolution programme that once again localised politics and reduced it to the exercise of political brokerage at the expense of the formulation of policies and legislation. Under Musharraf many of the leaders of the PML-N and the PPP were exiled, imprisoned and disqualified from elections. He created a parliamentary facade to his rule through the establishment of the Pakistan Muslim League Quaid-i-Azam (PML-Q), which was forged principally through defections from

---

[10] It is widely perceived among the general public that meddling by the military and the intelligence services also played a role in destabilising civilian governments during this period.

Nawaz Sharif's PML-N and to a lesser extent from the PPP. The extent of defections from the PML-N was such that 'On several occasions MNAs and Members of the Provincial Assembly (MPAs) have had to virtually be put under lock and key to prevent them from being lured by lucrative promises into crossing the aisle. In the spring of 1993, for example, a situation arose that reached farcical proportions. After the National Assembly was dissolved and Nawaz Sharif's government was dismissed from office, he was forced to sequester his party's Punjab MPAs in Islamabad's luxurious Mariott Hotel for nearly a month for fear that with their patron no longer in office, they would look for a new patron' (ibid.: 204–05).

Although in many cases these defections simply reflected the time-honoured opportunism of political leaders, they also reflected an active policy on the part of Musharraf's government whereby political leaders were both cajoled and coerced into defecting. Political leaders were cajoled through promises of lucrative positions in government and coerced by threats of prosecution on the basis of both real and fabricated charges. In order to weaken political opponents and rig elections, the judiciary was widely manipulated through the posting of pliable judges and the transfer and dismissal of judges who showed excessive independence. In addition, the National Accountability Bureau, though ostensibly created as anti-corruption watchdog, was in practice selectively used to prosecute political opponents. Although in all likelihood a great number of politicians could have been liable to prosecution on real corruption charges, only those who were in the opposition were prosecuted and those who decided to defect to the ruling coalition were exonerated of all charges.[11] Other tactics used in order to encourage defections included harassment and physical intimidation by the police and by the ISI. Finally, as will be shown in a later chapter, through his devolution programme General Musharraf was able to create a constituency that allowed him to bypass opposition provincial politicians and mobilise support for a rigged referendum, extending his term for five years based upon a 95 per cent approval rating.

In 2008 General Musharraf was forced to resign following electoral defeat at the hands of the PPP. This followed Benazir Bhutto's murder in

---

[11] Faisal Saleh Hayat, for example, an influential landlord in the central Punjabi district of Jhang, had been a major defaulter on bank loans and General Musharraf's National Accountability Bureau prosecuted him until all charges were dropped when he accepted to join the government in order to be made interior minister. See the International Crisis Group (ICG) (2004a) report on Pakistan's judiciary for an account of this case and of how the judiciary has been used by successive governments in Pakistan to prosecute political opponents.

September 2007, years of agitation by the judiciary – seeking greater judicial independence after years of meddling by the Musharraf government – and by the Alliance for the Restoration of Democracy, as well as years of growing Taliban insurgency in the North West Frontier Province. However stacked courts, a partial caretaker government, a subservient Electoral Commission of Pakistan, a gagged media, curbs on political mobilisation and association, and interventions by the security agencies had all indicated that Musharraf would retain power by once again holding rigged elections. But after Bhutto's murder – and the widespread belief that Musharraf was either directly or indirectly responsible for it – blatant rigging would have provoked violent protests so the Musharraf government restricted itself to selective rigging. The PPP eventually took power nationally and the PML-N regained power in the Punjab, thereby placing the PPP in a similar situation to the one it faced in 1988 when Benazir Bhutto was Prime Minister of Pakistan and Nawaz Sharif was the Punjab's Chief Minister. Despite serious governance issues, inflation, electricity shortages, and an ever growing militant threat, Zardari's PPP government was the first elected government in Pakistan to complete its term in office. During this period Pakistan's elected representatives enacted tangible measures to consolidate democracy among which was making the appointment of the Chief Election Commissioner more transparent and subject to parliamentary oversight – instead of the president handpicking him as had been the case historically. Thanks to this, the parliamentary elections of 2013 were largely free and fair and the PML-N regained power both nationally and in the Punjab. However it is still too early to judge whether the military will allow democracy to consolidate or whether it will revert to direct interventions in the country's political processes.

This brief overview of the political history of Pakistan has set the larger context within which local power structures and the landed elites became entrenched within the country's political system at the expense of the majority. It was argued that Zulfikar Ali Bhutto's democratic rise to power in 1970 briefly succeeded in overcoming parochial clan based and factional political allegiances through his appeal to national issues that were significant to the masses. However, the military takeover by General Zia-ul Haq, by banning political parties and by once again diverting the attention of politicians away from national to purely local issues, reversed whatever progress had been achieved towards popular empowerment. As a result, the vital national issues of land and homestead reform – as well as the reform of local governance structures, including the police and the judiciary – were never seriously addressed. As chapter five will demonstrate, contrary to its stated goals of reforming local governance structures and empowering the masses, General Musharraf's devolution programme was in fact an attempt to maintain the status quo by once again localising politics and preventing the consolidation of large scale political opposition, in the same manner that General Ayub Khan and General Zia-ul Haq had done in the past. This meant that rural notables were able to extend their stranglehold over rural areas. In turn the military often justified its repeated political interventions by pointing at the venality of these politicians; ostensibly overlooking the fact that the military was itself guilty of entrenching them in the country's political system. For most of Pakistan's short history, this generated a self-perpetuating cycle whereby civilian governments characterised by the factional politics of patronage were ousted by military governments who further entrenched factional politics by stunting political development. Whether the military will allow the country's political institutions to develop will determine whether Pakistan is condemned to perpetuating this cycle.

Gatekeeping as accumulation and domination

While a number of studies referred to in the preceding section have shown how central government policies served to entrench traditional power structures and prevent the emergence of more inclusive, popular political movements, few have given detailed accounts of the local mechanisms through which these power structures were reproduced. Matthew Nelson is one of the few authors to examine how government policies and legislation are systematically subverted by local power structures but his work doesn't examine how the subversion and appropriation of state programmes and resources reproduces these power structures. In this section I show how the private appropriation of state resources by an extended kinship group was the central mechanism through which dominance was reproduced. This not only consolidated the political and economic clout of the extended Gondal clan but also disempowered subordinate classes by making them dependent on the Gondals for access to state resources. I argue that clientelistic ties between landlords and the landless prevented the emergence of political unity among the latter.

The Gondals' compact clan structure, their numerical preponderance in the area and their landownership enabled them to obtain parliamentary seats in the provincial and national assemblies and to replace the aristocratic (*ashrafi*) Makhdooms as the principal political force in the region in the 1970s. The latter's political influence had largely derived from the fact that they owned over 1000 acres of land but their small numbers in the area made it difficult for them to compete in elections against the far more numerous and forceful Gondals. Moreover the Gondals were able to ride the popular wave of support for the anti-feudal, socialist rhetoric of

Zulfikar Ali Bhutto's Pakistan People's Party directed against landlords like the Makhdooms.

Chowdri Nawaz Ali Gondal, the eldest of five siblings, owned 70 acres of prime canal irrigated land at the time when he was the first Gondal to become involved in parliamentary politics. He had been the second Gondal after his father to obtain further education and to qualify as a lawyer. Unlike his father however, he completed his education in jail because he was implicated in the murder of a fellow student at Sargodha University. Law had been an ideal entry route into politics in a situation where clients most commonly approached their leaders to help them with land disputes and other cases involving the local police station (*thana*) and the local courts with their associated lawyers' offices (*kacheri*) (see Nelson 2011). The murder and his effective and astute handling of cases soon established Chowdri Nawaz Ali as an effective local power broker and he was eventually brought into the PPP by local party leaders. A member of the influential aristocratic family (mentioned in the introduction), who at the time was still a king-maker in the district, offered Chowdri Nawaz Ali a party ticket to run for a provincial assembly seat.

When I first met him in 2004, Chowdri Nawaz Ali claimed to still be a 'socialist' even though it had been nearly three decades since he had abandoned the PPP for Nawaz Sharif's conservative, pro-business party. When I asked him how a big feudal landlord like himself could be a socialist he corrected me, telling me that he was a member of the middle class (*darmian tabka*), not a feudal. He argued that although there were still plenty of 'feudal' politicians who owned hundreds, if not thousands of acres of land, obtaining political office no longer simply depended on status and land-ownership; to obtain it politicians had to do 'people's work' (*logon ka kaam*) by delivering patronage, jobs and infrastructure.

As this section will illustrate, it would be misleading to assume Chowdri Nawaz Ali's claims to mean that Pakistan has undergone political and economic developments that have made its state more responsive to pressures from below (as has been claimed to be the case in India by authors such as Inbanathan and Gopalappa 2002; Manor 2004). Moreover my findings contradict the arguments of development optimists such as Oliver Mendelsohn (1993) according to whom economic and political developments have diminished the power of South Asian landed elites to such an extent that it no longer makes sense to talk about dominant castes (Srinivas 1987). Following authors such as Pattenden (2011) and Jeffrey (2001) I show how formal democracy has in fact allowed landlords to consolidate their dominance at the expense of the poor.

Following Zia-ul Haq's military takeover and a stint in jail for being in the PPP, Chowdri Nawaz Ali opportunistically joined the new ruling coalition despite the fact that its ideology was diametrically opposed to the PPPs populist socialism. Over the years that followed, this shift in political allegiance proved highly beneficial to the political and economic fortunes of leading Gondals. From 1985 onwards both Chowdri Nawaz Ali and his younger brother Chowdri Mazhar Ali were to benefit greatly from the unprecedented levels of patronage that characterised Nawaz Sharif's several terms in power – both as Punjab Chief Minister and as Prime Minister. They and their allies benefitted from the fact that Nawaz Sharif filled thousands of government jobs with his supporters. Moreover Nawaz Sharif disbursed funds and granted licences to supporters and also appointed hundreds of loyalist police officers to help his supporters consolidate their local power and harass their rivals.[12] Chowdri Mazhar Ali first became a provincial minister in 1985 when General Zia was still in power and when Nawaz Sharif was Chief Minister of the Punjab. Following the death of General Zia, and after a two-year period with Benazir Bhutto as Prime Minister, Chowdri Mazhar Ali became an MPA during Nawaz Sharif's two terms as Prime Minister, from 1990 to 1993 and 1997 to 1999. In the meantime, Chowdri Nawaz Ali who had started his political career with the PPP also joined Nawaz Sharif's PML-N and came to play a prominent role in its district organisation.

Over almost 14 years the Gondals intermittently benefitted from substantial amounts of patronage when Nawaz Sharif was in power. This provided them with a wide variety of opportunities to further their personal economic and political standing. This is most dramatically illustrated by the fact that by 1999 Chowdri Mazhar Ali owned almost 400 acres of land and several Pakistan State Oil (PSO) petrol stations throughout Sargodha District, whereas when he began his political career in 1985 he had owned no more than 70 acres of land. Similarly Chowdri Abdullah Gondal – the son of the lambardar Ahmed Rasool Gondal, who shared a paternal great-great-grandfather with Chowdri Mazhar Ali – dramatically increased the amount of land he owned from around 20 to also over 400 acres. He did so by making money in the sand business, after obtaining a licence for its extraction from Chowdri Mazhar Ali, through heroin trafficking and by

---

[12] Wilder reports that during his time in power Nawaz Sharif was able to fill thousands of government jobs with his supporters: 'he appointed hundreds of loyalist police officers, particularly into the lucrative positions of Assistant Sub-Inspectors and Station Head Officers. This was especially significant as the police play a central political role in Pakistan because of their ability to selectively apply laws in order to harass opponents or to turn a blind eye to the misdeeds of political allies' (Wilder 1999: 139).

grabbing land from the once politically influential Makhdooms as well as from poor peasants.

During his three terms in office Mazhar Ali Gondal obtained a variety of lucrative contracts and a steady flow of development funds for his factional constituency and for his biraderi. For example, during his latest term in office, from 1997 to 1999, Chowdri Mazhar Ali obtained contracts for the construction of stretches of the motorway from Lahore to Islamabad that was being engineered by the South Korean Daewoo Company. He appears to have not only personally taken advantage of these contracts but to have granted contracts to his relatives including his younger brother, who became extensively involved in the project, and other supporters. His younger brother was even somehow able to place one of his domestic servants onto the payroll of Daewoo without the servant ever having to actually work for the company. According to the servant in question this arrangement basically allowed his master to have a servant for free.[13] Chowdri Mazhar Ali's younger brother, his nephews and cousins and other allied Jats, were also granted several local road construction projects. The state of many of these local roads only a few years after completion suggests that the contractors in charge of their construction spent far below what they had invoiced. The metal road running alongside the canal near Bek Sagrana had only been built seven years earlier, but in 2004 it was already badly affected by potholes and large parts of it had collapsed into the canal. Additionally, Chowdri Mazhar Ali was granted several franchises to build PSO petrol stations along the main roads leading into Sargodha.

During Chowdri Mazhar Ali's time in office the village of Bek Sagrana also received a large number of development projects which Gondals obtained contracts to build. The village gained a post office, a water tower, a basic health unit, a new school, and a community centre. Of these, neither the water tower nor the post office was ever used and the basic health unit wasn't operating. The water tower was part of a scheme originally destined to connect households to a water-supply system in areas where there were problems with water scarcity and salinity. Bek Sagrana suffered from neither of those problems, and most people freely obtained fresh water by the means of a hand-pump (*nalka*) located in the courtyard of their house. No one in the village opted to get connected to the new water system because they saw no reason to opt for a system in which they would have to pay monthly water bills. The post office was never operational and was used as the house of one of Mazhar Ali's ex-employees.

---

[13] The servant even told me that because the Daewoo salary was quite high, his master kept part of it for himself.

The basic health unit, like many others I visited throughout the district, wasn't properly operational. The unit included a hospital building with two operating rooms for minor operations such as appendicitis and childbirth, and a medical dispensary. It also had housing for hospital staff and a large two-storied residence for the doctor. The latter, a wealthy Gondal chowdri who had managed to get a position in the basic health unit of his own village thanks to the patronage of Chowdri Mazhar Ali, never attended to his duties there and only occasionally went to enter false entries into the attendance register.[14] The residence of the basic health unit was occupied by his maternal cousin, who was also Chowdri Mazhar Ali's nephew. A few of the rooms meant for hospital staff were occupied by the former's servants, one of whom used another of the rooms to keep broiler chickens which he sold in the village. The clerk in charge of the medical dispensary allegedly made money selling off the basic health unit's medical supplies. As a result villagers continued to rely almost exclusively on local healers (*hakeems*) and on others with highly dubious qualifications as practitioners of allopathic medicine.[15]

Although it is true that poor voters might occasionally obtain cash during elections, the intercession of a Gondal to resolve a court case, treatment in a public hospital, or even a government job, such brokerage didn't compensate for their loss of access to public resources through corruption. In other words what the preceding paragraphs illustrate is that the appropriation of the spoils of power by politicians didn't, as Lieven (2011) and Lyon (2004) claim, make state resources more accessible to the common man. Instead of access to health care, communal spaces, a decent infrastructure, and educational system and justice, most villagers just got cash payments in exchange for their votes; and the money rarely lasted them for more than a week. The lucky ones on the 'bright side' of political society, might obtain low ranking government jobs but did so in ways that undermined public service delivery and that were therefore detrimental to most of the population. In other words the dark side of political society was far more significant than its bright side which only benefitted the powerful and the well connected.

Moreover although, as will be shown below, some landless villagers did obtain government jobs thanks to the Gondals the overall evidence

---

[14] Chowdri Nawaz Ali was married to the doctor's maternal aunt.

[15] This sort of thing was very common in the public health sector. Most Basic Health Units that I visited in rural areas weren't fully operational and doctors meant to work in them only came to collect their pay once a month. See Asian Development Bank (2005) for an account of poorly motivated and absent staff in hospitals as well as in schools.

indicates that in fact the Gondals sought to prevent villagers from being empowered through jobs and through improved access to education and other public resources. Villagers repeatedly told me that the Gondals didn't want them to get good jobs and improve their lot in life because this would deprive them of young men to act as gunmen for them; whenever the Gondals had land disputes, or when they contested elections and needed to fight their rivals, they demanded that their tenants and labourers pick up arms to fight for them. They also repeatedly told me that whenever the Gondals noticed ordinary villagers wearing nice clothes, or talking on a new mobile phone, they became annoyed because they felt that they were losing their slaves (*ghulaam*) and responded by making fun of them and humiliating them.

The Gondals claimed that kammis nowadays wanted to be like kings (*badshah*) and that they no longer wanted to do respectable hard work. Instead they wanted to lounge about wearing 'flashy clothes' and make money 'the easy way' by running shops or by stealing and bootlegging. Although the Gondals didn't actively obstruct the spread of education like Southern Punjabi landlords were alleged to do, they didn't encourage it either.[16] The only member of the sweeper caste (*Mussalli*) to have obtained his matric also told me that the Gondals couldn't bear the fact that he was educated after one of them had mockingly asked him if he was a big man (*bara admi*) now that he spoke English with me, a foreigner. Confirming their distaste for popular education, several Gondals told me that education only served to implant unrealistic job expectations in people's minds and made it more difficult for them to obtain labourers. They complained that nowadays the only way for them to lure labourers was by providing them with small consumption loans, and in some cases larger loans to cover wedding or medical expenses (see Martin 2009). Moreover they didn't think that members of menial clans (*kammis*) were likely to reap a great deal of benefit from education anyway because they were inherently incapable of using their reason and of doing hard work.

As a result the Gondals didn't show much interest in fostering secular education and focused instead on the inculcation of morality through Islamic education. Instead of sending their child servants, working to repay debts incurred by their parents to cover wedding or medical expenses, they gave them some basic homeschooling and taught them how to read the Holy Qu'ran. They also built mosques and paid preachers to teach village

---

[16] A wealthy landlord and hereditary saint (*pir*) from the Southern Punjabi district once boasted that he didn't let schools operate in his region because they created unrealistically high expectations among labourers.

children about Islam but made little effort to help improve the quality of education being delivered in the village primary school. One school teacher who was from the village told me that because the Gondals increasingly educated their children in town they no longer had any incentive to make sure that the village school provided students with a reasonable education. Previous generations of Gondals had done much of their primary schooling in the village and as a result their parents had ensured that teachers turned up to do their job and that they taught properly when they were around. Now that the Gondals weren't around, there was no longer anyone with power and authority that was willing to ensure that teachers did their job. The result was high levels of absenteeism which teachers got away with by placing false entries into their attendance registers. Moreover even when they were around teachers spent significant amounts of time chatting with each other, drinking tea and performing their prayers in the village mosque. They often disrupted children's studies by sending them on errands to buy cigarettes and biscuits for them. When the school teacher from the village asked some Gondal landlords to personally impose fines on absentee teachers there was an initial show of goodwill but nothing came of it. Instead those very same Gondals soon started recruiting school children to run errands and even to gather fodder for their livestock. For all these reasons the literacy rate in Bek Sagrana at the time of the 1998 census was a paltry 18.5 per cent.

The Gondals were also largely averse to helping the few kammis with the requisite educational qualifications to obtain low-ranking jobs as clerks or guards (*chowkidars*). Chowdri Mazhar Ali, who was thought to be more forward thinking than other Gondals, had sought and managed to obtain low-ranking government jobs as guards (*chowkidars*) for four smallholders and three kammis during his time as an MPA but had faced resistance from his relatives who worried that that this would deprive them of labourers. In one widely discussed case, Chowdri Mazhar Ali obtained a class four job in the Ministry of Agriculture for a blacksmith (*Lohar*), but because his nephew wanted the Lohar to continue working as a labour supervisor in his brick kiln he was asked to withdraw the offer. Moreover, in half of the cases where Chowdri Mazhar Ali had obtained government jobs for poorer villagers he had done so to obtain free servants for himself and his close kin; the villagers in question were on the payroll of the state but worked for Gondals landlords as domestic servants, labourers and overseers. Thus one Rajput smallholder and a kammi breadmaker (*machi*) were on the payroll of the ministry of agriculture but actually worked as overseers on Chowdri Mazhar Ali's estate. The two smallholders and one kammi that were employed as chowkidars on the premises of the Basic Health Unit (BHU) spent a great deal of their time serving one of Chowdri Mazhar Ali's nephews who used the bungalow meant for the doctor employed at the BHU as a weekend retreat.[17] Finally a landless member of the Lurka biraderi had been on the Korean company Daewoo's payroll for several years while actually working as a driver and farm servant for Chowdri Mazhar Ali's younger brother Chowdri Arif.

Therefore some people other than the Gondals did benefit from the rise of popular politics, but they did so in ways that were at the expense of public service delivery and of the political unity of the poor. Since people with government jobs were actually working as private servants for the Gondals they at best only partially fulfilled their duties as government servants and therefore undermined public service delivery. Moreover the promise of patronage meant that, as Alavi (1971; 1972a) and Mohmand (2008) have shown, the landless sought to attach themselves with powerful patrons rather than to organise themselves politically in order to defend their interests. Instead of voting together as a block to secure their common interests they voted individually in order to please particular Gondal landlords and to try and secure their patronage.[18]

It is also true that the Gondals frequently provided people with patronage in the form of access to both formal and informal channels of dispute resolution. However like the forms of patronage discussed above, this tended to help them consolidate their own political and economic clout and tended to benefit their close allies – usually fellow Jats. Moreover just as patronage undermined public service delivery, so patronage encouraged the growth of crime and undermined law and order and meant that neither people or their possessions were ever safe. Gondal politicians frequently interceded in the court system and with the police on behalf of criminals who worked for them as toughs. Because they protected buffalo thieves people had to sleep next to their livestock at night. Likewise because they protected heroin traffickers, the village became a converging point for heroin addicts who stole anything that crossed their path. Thus throughout my

---

[17] The bungalow was a sort of bachelor pad where he invited friends to drink. He once even used it to bring over some dancing girls from Sargodha.

[18] My view of patronage sharply contrasts with that of Pettigrew – working in Indian East Punjab during the 1960s – who claims that clients that 'are not those to whom favours are done, the victims of a system of charity that makes piecemeal adjustments to the inequalities of life. They have the capability to reciprocate for what they receive from a patron who is different from them only at that one point in time when they ask a favour' (Pettigrew 1975: 20). See especially Gellner (1977) for a discussion of patronage and of how patronage ideologies falsely claim to be based on reciprocity.

stay I heard of people's bicycles, clothes and agricultural implements being stolen by heroin addicts.

In addition to encouraging criminality the exercise of patronage in the area of dispute resolution undermined impartiality in the delivery of justice in both formal and informal dispute resolution forums (see Chaudhary 1999) and meant that people with the right connections could get away with a wide range of illicit and socially unacceptable forms of behaviour. In one instance, for example, a Lurka who was widely known to beat his wife when he was drunk managed to escape prosecution because he had good relations with the head of one of the village factions. His brother-in-law spent a great deal of time and money in order to file a police report (a First Information Report) against him but he was eventually forced to withdraw it when he was threatened with eviction by landlords allied with Chowdri Abdullah on whose land he had built his house. Like many other villagers he complained that the Gondals were only concerned with consolidating their power and had absolutely no interest in justice (*insaaf*). As a consequence villagers spent a great deal of time trying to ingratiate themselves with the landlords, and some even went to the extent of snitching on each other to gain their favour. Thus whenever a theft took place there was never a shortage of villagers willing to convey information – either true of fabricated – to the Gondals about who performed it. This gave rise to widespread mistrust and suspicion of the village poor towards each other and to claims that Bek Sagrana was a bad place because it was full of snitches (*chugl khor*). From the moment of my arrival in the field I was repeatedly told that villagers couldn't be trusted and that they were for the most part rotten (*kharaab*) and fraudulent (*do nambar*).

Conclusion

While much of the literature on Pakistan examines higher level political and institutional events and processes, this chapter followed the anthropological tradition of studying the 'everday state' (see Fuller and Benei 2001); the state as ordinary people experience it on a daily basis. While higher level political events and processes are undoubtedly important, what ultimately matters for the students of both sociology and of government is how and whether these impact people's lives. Despite focusing on the everyday state however, the analysis presented is not typically anthropological because it doesn't invoke 'traditional' cultural norms and social structures to explain how the Pakistan state works. It shows how the Pakistani state's failure to deliver basic services – such as health care and the rule of law – relates to the historical collaboration between authoritarian military regimes and the landed classes, rather than merely to how local kinship structures undermine impersonal legal and bureaucratic norms – as both Nelson (2011) and Lieven (2011) claim.

In this chapter I argued that the consolidation of the landed class, thanks to repeated military interventions, contributed to depriving a large part of the rural electorate of access to basic state services. These repeated interventions achieved this by forestalling the emergence of the large scale political movements, or even the rights based movements, that could have challenged the landed class and/or forced it to become accountable and to share more of the state resources that it appropriated for itself. I also suggested that military interventions even further reduced political accountability by arbitrarily empowering loyal politicians – rather than allowing the people to empower them – and bending state institutions on their behalf. In this way notoriously corrupt and even criminal politicians were allowed to amass fortunes and, in some instances, protected from prosecution by the criminal justice system and by anti-corruption bodies. Conversely opposition politicians were prosecuted, harassed and jailed, their businesses were closed and their land was grabbed. Military rule facilitated the process whereby politicians could use their upward links with higher level politicians to bend state institutions – including the judiciary – in order to serve their own narrow interests.

While Bek Sagrana may not be representative of the Punjab as a whole, it provides us with clues as to the causes for poor public service delivery in the state, and beyond. The first cause, discussed above, is despotism – in the sense of arbitrary authoritarian rule – starting at national level and extending through the provinces and down to the villages. Despotism has undermined governance throughout Pakistan. The second cause I propose is inequality, a factor more specific to Bek Sagrana but by no means unique to it.

Bek Sagrana was a reputedly poor and backward village, widely perceived to be populated by illiterate Mussalli cattle thieves and their Gondal feudal patrons. Wealthier villagers believed that a particularly low literacy rate – among the lowest in the area – was the reason for backwardness and criminality in the village. Illiteracy and ignorance of the basic tenets of Islam were allegedly the reasons why Mussallis trafficked drugs, distilled alcohol and stole buffalos. However literacy and knowledge of the basic tenets of Islam didn't stop wealthy landlords from providing cover for these activities. More convincingly, people argued that illiteracy made people unaware of their rights and therefore unlikely to complain when the state didn't deliver basic health care and education.

However illiteracy was but one facet of the relative deprivation that prevented kammis from demanding their rights and pressing for improved government services. The broader problem was that kammis were too poor and fragmented to challenge the powerful Gondals and to make demands from them. As a later chapter will illustrate, when the Gondals didn't automatically command kammi votes they either purchased them with patronage and/or cash or simply coerced them. The Gondals could buy votes because kammis needed to satisfy pressing immediate wants – such as the want for food – and couldn't afford to wait for longer term and less tangible benefits such as improved universal education and health care. Thus the combination of poverty and inequality prevented the electoral process from producing leaders that were broadly accountable and likely to improve public service delivery in their constituencies.

Poor governance in Bek Sagrana was therefore born of a combination of personalised authoritarian rule at national level and of local conditions of poverty and inequality. Personalised authoritarianism operates throughout Pakistan while poverty and inequality characterise much of rural and urban Pakistan. However this doesn't mean that all villages with similar levels of poverty and inequality as Bek Sagrana will be like it because leaders differ in their integrity and desire to benefit their constituents. Thus while many broadly similar villages I visited throughout Sargodha district and beyond had equally dysfunctional basic health units and schools, in a minority of them these worked well, thanks to the oversight of benevolent and paternalistic leaders. Further research would however be necessary to establish what type of leadership structures made it possible for leaders like these to resist close supporters trying to get them to subvert state institutions on their behalf.

4
THE ENEMY OF MY ENEMY IS MY FRIEND

Biraderi and factionalism

... because nearly every village is divided into factions, and these are often linked into larger factional groupings, their cumulative effect can be great. This poses a challenge to political parties seeking to increase their support in rural constituencies. No matter which candidate is selected, strong opposition is virtually guaranteed by rival factions.

(Wilder 1999: 177)

In every village there were generally faction feuds raging. Active minds were busy ... devising plans for dishing the other side, detaching members from it, and generally putting their own side ahead.

(Carstairs 1912)

In the previous chapter I suggested that authoritarian central governments as well as local power structures sustained by poverty and inequality were responsible for poor health care, education, justice, and security. I argued that authoritarian central governments were responsible for this to the extent that they prevented the emergence of popular political movements seeking to challenge traditional power structures and to the extent that they undermined the impartiality of state institutions to consolidate their power.

I also indicated that by localising politics and placing restrictions on political parties, military governments were able to consolidate their power and fragment popular political opposition. Localising politics and making policies and passing legislation without consulting elected assemblies diverted politicians' attention away from policy making and legislation and left them to focus all of their energies on local patronage politics and on frequently petty but violent power struggles. Moreover financial restrictions on political parties increased the dependence of political parties on the private wealth, connections and ability to mobilise force of their candidates to contest and win elections.

By localising politics successive military regimes ensured that personalised ties and private feuds rather than programmatic political agendas remained central in determining the political allegiances of both politicians and their followers. Thus Ayesha Jalal (1995), and others (see Burki 1980; Rais 1985; Waseem 1994) have argued that kinship ties, in the form of the extended lineage (biraderi), regained their political force during the Zia regime. Ayesha Jalal writes that in a 'country where parties had never managed to strike roots, the Zia regime's systematic campaign to discredit politicians and politics gave renewed significance to the old personalized networks of biraderi or clan-based ties' (Jalal 1995: 105). This chapter examines how these local ties of allegiance structure political competition and conflict, and how they determine the party membership of politicians.

In Pakistan both academic and popular discourses tend to emphasise the idea that primordial loyalties override ideological and party allegiances. Thus, it is frequently argued that the assumption behind a democratic system, whereby individuals freely make their own choices in all spheres of life including politics, does not apply in a context where primordial loyalties give precedence to the group over the individual. Ullah, for example, made the following claim:

> The first important fact about the village life is that it is an aggregate of individuals. In fact, the real individual in the sense of Western Urban society does not exist in the village. He is an inalienable part of multiple groups which completely overshadow his individuality . . . The first and most important group for the individual is the family which makes for him the major decisions of life . . . Next to the family comes the '*Baradari*' [sic] group.
> (Ullah 1963: 50–51)[1]

Despite the fact that several authors invoke kinship to explain the failure of political parties to establish roots in Pakistan, few really examine how it actually determines people's political allegiances.

As Wilder's quote above indicates, extended biraderis – like the Gondal biraderi – tend to be divided into factions and rarely act as a united political unit by for example voting as a block. Unlike in segmentary systems (most commonly used by scholars to analyse Middle Eastern tribal society)[2] where

---

[1] See also Ullah (1958).
[2] See Gellner (1969) for a classical account of a segmentary social system in the Atlas Mountains of Morocco.

feuding agnates leave their rivalries aside and unite in the face of external aggression, the Gondals – like Fredrik Barth's Khans – frequently aligned themselves with outsiders against their own clan members and even their close kin. In the classical segmentary model (most commonly used by scholars to analyse Middle Eastern tribal society) the presence of shared interests in a joint estate, together with a sense of honour attached to the tribe, cause opposed segments at one level of segmentation to unite at a higher level of segmentation when faced with an external threat. In such a scenario it might be expected that, despite their internal quarrels, Gondals would always support a Gondal politician over and above a politician belonging to another biraderi since to support an outsider would undermine the interests of the biraderi as a whole.

The empirical evidence presented in this chapter does not support the view that the extended biraderi is the principal building block of political activity in the Punjab. Instead it will be argued that factions characterised by 'a vertical structure of power which cross-cuts caste and class divisions' (Brass 1965: 236), and which are based upon a loose coalition of individuals united by common enmities and structured around a closely knit kinship based core, are the principal building blocks of political activity and allegiance.[3] Following Barth in 'Segmentary Opposition and the Theory of Games' (1981), this chapter presents the argument that, even if it is possible to talk of members of an extended biraderi sharing in a joint estate, this possibility 'need not imply a community of interests, and may in fact imply an overriding opposition of interests which inhibits the emergence of corporate unity' (Barth 1981: 81). Barth argues that amongst the Swat Pathans it is the overriding opposition between patrilateral cousins (*tarbur*), who almost invariably compete over the possession of land and the control of client groups implied by it, which inhibits the emergence of the corporate unity of the Yusufzai Pathans. According to Barth's model, a Pakhtun's 'political activities are directed at gaining an advantage over his agnatic rivals, as only through their defeat can he achieve his own aggrandizement' (Barth 1981: 67). To this end agnates make political alliances with distant collaterals against each other and follow the principle that the enemy of their enemy is their friend. Barth argues that the principal reason why Pakhtuns can afford to oppose their agnates is that, unlike in other lineage systems, their agnates do not form the bulk of their supporters. Instead the bulk of a Pakhtun's supporters are gained from within the ranks of dependent client groups, including artisans and tenants, whose loyalties

---

[3] Washbrook (1976), Weiner (1967), Bailey (2001), Lewis (1958), Nicholas (1965), and Brass (1965), have all used the concept of factions in their political analyses.

Pakhtun patrons win through the distribution of largesse made possible by their ownership of the land. The overriding importance of the rivalry between agnates, and the fact that leaders make alliances with the enemies of their enemies, encourage the regional formation of a two-block system cutting across extended biraderi ties.

This chapter shows that Barth's model is applicable to the political coalitions made by the chowdris of the Punjab, where agnatic rivalry often obliterates the unity of the blood group. Contrary to Lindholm (1982) I argue that the unity of the blood group is not necessarily activated in cases of revenge, and that factional leaders do in fact frequently support the enemies and killers of their close agnates. By demonstrating that chowdris, particularly powerful ones, often do align themselves politically with distant kinsmen and non-kinsmen against more closely related kin, I also question Alavi's view that 'it is only rarely that a household finds itself with no option but to align itself with more distant kinsmen and pitted against closer relatives' (Alavi 1972a: 18).

I argue that the political alliances of chowdris in contemporary Pakistan are not structured by mechanical lineage solidarity but by both the largely strategic choices of leaders and followers and by the rivalry and enmity between agnates, who compete over land and political resources. In terms of political party alliance what this means is that politicians attach themselves to the political parties with which the enemies of their enemies are aligned. In other words their choice of political party is largely unrelated to ideological considerations or even class interests.

However, the analysis presented differs from Barth's in that the role of kinship is not merely a negative one (in the sense that it structures enmity rather than creating lineage solidarity) but that the individuals who comprise the core of factions are often held together by multiple, overlapping kinship ties. Because kinship amongst West Punjabi Muslims is structured around preferential cousin marriage[4] this results in 'compact, tightly organised and self contained but, at the same time, small and localized biraderi groups' (Alavi 1972: 26). The material illustrates that it is such 'tightly organised and self-contained' kinship groups, united by affinal ties, which

---

[4] According to Alavi, the Punjabi Muslim kinship system is structured on the principle of preferential patrilateral parallel cousin marriage. I have gathered insufficient evidence to confirm this but it appears that cousin marriage is the norm and that a great deal of prestige and honour (*izzat*) is attached to endogamy. Prestige is particularly attached to keeping daughters within the biraderi, and wealthy peasants and landlords often boast about how they take girls but do not give them (see Alavi 1972a: 6).

form the core of political factions amongst chowdris.[5] Additionally, the chapter shows that agnatic rivalry does not stem only from competition over the possession of land. Whereas Barth characterises Swat as having an acephalous political system, the political system of the Pakistani Punjab is clearly not acephalous, and the Pakistani state plays an important role in structuring political competition and conflict.[6] Thus in the present day Punjab it is principally for control over the state apparatus that landlords fight. Although land is still important, landlords recognise that their ability to protect it from the encroachment of rivals, as well as their capacity to acquire more land, partly hinges upon their capacity to mobilise both the police and the courts in case of a land dispute. Political office and positions in the state bureaucracy are also well known for facilitating the acquisition of land through the multiple possibilities for enrichment that they offer. This means that conflict between agnates often results from competition for votes during elections for various tiers of government, as well as from competition over control of state institutions such as the police and the judiciary.

Factional leadership and competition in Bek Sagrana

The Gondals seemed to focus much of their political energies on violent private feuds. Because political contests between agnates at village level were zero sum games – in which one leader's political and economic gain was another's loss – agnates spent a great deal of their time trying to keep each other down. These contests were zero sum games because land and political office – the two sources of power – were limited resources which people could only gain at each other's expense. Whenever one faction happened to be aligned with the ruling coalition it not only ensured that its members gained access to state resources but also frequently that rivals were deprived of them and even actively victimised and persecuted to ensure that they remained weak. Thus when Chowdri Mazhar Ali and

---

[5] Alavi (1972a) and Eglar (1960) both describe the institution of *vartan bhanji*, referring to the exchange of gifts and favours, particularly at times of marriage, which serves to integrate households. I do not explore these in this chapter because they are not strictly central to the discussion and also because most of these exchanges centre on women to whom I had no access. Also see Mundy (1995) where the author similarly argues that households rather than the tribes are the core of political organisation in Yemen.

[6] Barth's characterisation of the Swat political structure as acephalous is in any case problematic, as Ahmed (1976) argues, since it failed to take into account the colonial administration's crucial role in propping up local leaders.

Chowdri Nawaz Ali were out of office during the nine years that Nawaz Sharif was exiled under the Musharraf government, Chowdri Abdullah – their principal rival in Bek Sagrana – was able to use his contacts within the ruling coalition to aggressively expand his wealth in terms of land and real estate. Chowdri Abdullah also managed to entangle them and their followers in a variety of court cases, including a fabricated one in which they were accused of harbouring terrorists. They also ensured that members of Chowdri Mazhar Ali's faction in the village were denied access to a phone line being installed by the government. However when General Musharraf's ruling coalition was ousted in 2008 the tables turned on Chowdri Abdullah. He ceased to accumulate riches and land and Chowdri Mazhar Ali even managed to get him jailed for illegally cultivating rice on the strip of government land next to the metalled road that ran past the village.

Chowdri Abdullah Gondal and Chowdri Nawaz Ali and his siblings shared a great-great-grandfather known as Kala Gondal. It was only in the previous generation that the descendants of Kala Gondal had become divided into rival factions that came to be known as the Lambar Ke and Ghulam Baksh Ke factions. The Lambar ke faction was named after Chowdri Abdullah's father who had been a local revenue collector (*lambardar*) while the Ghulam Baksh Ke faction had simply been named after Chowdri Ghulam Baksh – the grandfather of the current factional leader. Despite their rivalry, members of both had often joined forces against outsiders – particularly the Makhdooms – until around the 1990s. Subsequently when the Makhdooms started losing their political clout there was less need for unity among the Gondals and factional rivalry intensified.

Chowdri Nawaz Ali and his younger brothers were aligned with the Ghulam Baksh Ke faction headed locally by Ghulam Baksh's grandson Chowdri Haq Nawaz Gondal. This core was composed of Gondals with multiple overlapping kinship ties and who spent a great deal of time visiting each other.[7] Although neither Chowdri Nawaz Ali nor his siblings lived in Bek Sagrana, they provided Chowdri Haq Nawaz with the patronage that was necessary for him to retain his influence as head of the faction at village level. On the other hand Chowdri Abdullah Gondal didn't have

---

[7] Chowdri Nawaz Ali and Chowdri Mazhar Ali were the paternal first cousins, as well as close affines, of Hajji Sahib. Hajji Sahib had married one of Chowdri Nawaz Ali's sisters and Chowdri Nawaz Ali and a younger sibling had married two of Haji Sahib's sisters. In the next generation Hajji Muhammad Hayat's eldest daughter had married Chowdri Nawaz Ali's only son.

close kinship ties to powerful politicians and his local influence had largely derived from his having three forceful younger siblings to support him and from strategic alliances with various provincial politicians who shared his enmity with Chowdri Mazhar Ali and Chowdri Nawaz Ali.[8]

While the rivalry between Ahmed Rasool – Chowdri Abdullah's father – and Ghulam Baksh had been tempered by their joint rivalry against the Makhdooms, there was no longer a common external enemy to unite the two factions in the generation that followed. Moreover the fact that Chowdri Abdullah's fortune soared dramatically reduced his reliance on fellow Gondals for political support. The eventual consequence of this was that the unity of the Gondal biraderi in Bek Sagrana was completely obliterated.

Elderly villagers fondly recalled a time when, despite rivalries and occasional land disputes, all of the Gondal village elders had sat together in the village men's house (darra) in the evenings in order to address village problems and disputes jointly. Four of the influential village chowdris at the time, including Ahmed Rasool and Ghulam Baksh, had shared a monthly rotation for covering the expenses of feeding guests and providing a hookah at the village darra. Even following the deaths of Ghulam Baksh and Ahmed Rasool both factions had cooperated as the Gondals challenged Makhdoom dominance in the area. During the 1970s and 1980s, Ahmed Rasool and his son had even supported Chowdri Nawaz Ali and his brother during elections. Subsequently, while Chowdri Mazhar Ali was in power he had provided various forms of patronage to Ahmed Rasool and later to his son Chowdri Abdullah. Most widely recalled was the occasion when Chowdri Mazhar Ali had managed to obtain Ahmed Rasool's release from jail. Ahmed Rasool, like many of his cousins, had been extensively involved in protecting buffalo thieves, an activity known as *rassa-giri*,[9] and had been jailed as a result of it during Zulfikar Ali Bhutto's second term in power. It was said that without Chowdri Nawaz Ali's intervention to obtain his release Ahmed Rasool would have been blinded as a result of

---

[8] Chowdri Abdullah happened to be married with one of Chowdri Nawaz Ali's half-sisters from his father's side. This, however, didn't help relations because Chowdri Nawaz Ali had an on-going land dispute with his half-siblings, including Chowdri Abdullah's wife.

[9] The term *rassa-gir* refers to those landowners who act as protectors for gangs, thugs and thieves. The term literally translates as 'he who holds the rope' and refers to the fact that landlords metaphorically extend a rope to those they protect, to pull them out of trouble.

police torture.[10] In the years that followed, Chowdri Mazhar Ali had also granted Chowdri Abdullah various contracts and had provided him with cover from the police for him to pursue various criminal activities with impunity. Chowdri Mazhar Ali told me personally that it was he who had granted Chowdri Abdullah his government licence for the extraction of sand for the entire district of Sargodha, and that it was in part thanks to this contract that Chowdri Abdullah's economic fortunes had soared.[11] Chowdri Mazhar Ali also claimed to have given considerable support to Chowdri Abdullah in his land disputes with the Makhdooms.

However, as the influence and wealth of Chowdri Abdullah and his three siblings grew they became increasingly assertive and started to challenge Chowdri Haq Nawaz and his patrons. Over the years Chowdri Abdullah Gondal and his three younger siblings aggressively expanded their local political and economic influence through a variety of forceful and criminal, methods. This led them to acquire a well-deserved reputation for being enterprising as well as belligerent and violent, to a degree that surpassed nearly all other Gondals in the region.

Rival leaders in Bek Sagrana

Like his father Ahmed Rasool, Chowdri Abdullah spent a great deal of his time cultivating contacts among influential local, regional and national-level officials in order to consolidate his political and economic position. His ceaseless networking, coupled with his ambitious and domineering personality, eventually turned Chowdri Abdullah into a highly effective local political figure. Thus, although Chowdri Abdullah (unlike Chowdri Nawaz Ali and Chowdri Mazhar Ali) never achieved political office above the lowest administrative unit, his wide-ranging personal contacts and friendships allowed him to become an influential power broker in the area. The force of his personality, the awe in which he and his siblings were locally held, and his many gunmen meant that he was greatly feared, but it also meant that most villagers believed him to be the local leader most likely to intervene effectively on their behalf in a dispute. If, for example, a smallholder who had good relations with Chowdri Abdullah had a land dispute with a

---

[10] Apparently the police officers had applied intense pressure to his eyes with their thumbs. Ahmed Rasool had been jailed during the time that Mustapha Khar, also known as the lion of the Punjab (Sher-e Punjab) and the subject of Tehmina Durrani's (1994) book *My Feudal Lord*, was Chief Minister of the province and had decided to crack down on buffalo theft.

[11] Chowdri Mazhar Ali's claim to have done so was confirmed to me by a number of villagers.

neighbour, the mere fact that Chowdri Abdullah supported him was usually enough to make his opponent withdraw. Chowdri Abdullah was reputed to have an uncontrollable temper and anyone who angered him was likely to face his wrath. Villagers told me that if he caught a servant stealing from him, or a villager stealing from his citrus orchards, he would go into violent fits of rage and would personally inflict beatings on the offender. As a result villagers who often freely picked citrus in other chowdris' orchards didn't dare do so in Chowdri Abdullah's orchards.[12]

Although Chowdri Abdullah was known for his temper and violent acts, it was Chowdri Rafiq, his youngest brother killed in a police encounter, who was the most notorious of the four male siblings. At the time of fieldwork in 2004 he had been lying dead for over seven years in a large domed tomb (*mazaar*) next to his family's farmhouse (dera). It was Rafiq Gondal who was credited with the introduction of heroin into the area of Bek Sagrana in the 1980s, when the drug started to flow copiously into Pakistan from Afghanistan. Some of his relatives even claimed that he had extended this business to Lahore. Villagers remembered Rafiq Gondal with awe, and stories about his reckless bravado, and his equally reckless generosity, abounded in the village. His personal gunmen recalled how, unlike most other chowdris, he had personally accompanied them to fight against his rivals and occupy their land. One gunman recalled how on one occasion Chowdri Rafiq personally set off in the middle of the night to find one of his client's stolen buffalos. When Chowdri Rafiq found the thief swimming across the irrigation canal with the buffalos, he jumped in and caught the thief with his own hands. The same gunman recalled how Chowdri Rafiq had once caught one of his local enemies, tied him down to a charpai, and

---

[12] Villagers also recounted an occasion when he ran a rickshaw into an irrigation canal because it had failed to give way to him quickly enough as he sped along the road in his large pickup truck. One villager who had been given accommodation by Chowdri Abdullah in an outbuilding next to his town house in Sargodha told me that Chowdri Abdullah had once rammed into the gates of the compound with his pickup because the *chowkidar* had been too slow in opening them. Yet another story related an occasion when Chowdri Abdullah had slapped his son in the face in front of a large assembly of villagers because the fact of his son's being taller than he was annoyed him. However, the most violent act committed by Chowdri Abdullah was described to me by his nephew. His nephew claimed that, following his younger brother's death in a police encounter in 1999, Chowdri Abdullah had taken revenge by going to the house of one of the police officers involved, shooting him, and burning down the house. In short, few people, other than the leaders of the Ghulam Baksh Ke faction, would dare to provoke the wrath of Chowdri Abdullah.

broken both of his legs with a brick. He also recalled how, when Chowdri Rafiq was feeling generous and pious, he would give out 1000-rupee notes to his gunmen and tell them to take a few weeks off work. Another man recalled how on one occasion, when Chowdri Rafiq had been listening to loud music in his car in the nearby market town, he had violently slapped a policeman who had ordered him to lower the volume. Faced with a heavily armed group of men the policeman had apparently meekly turned around and returned to his post.[13]

Forceful behaviour and a reputation for it were assets for local politicians, particularly for those operating below the provincial assembly level. This is because they were the ones who did much of the dirty work of delivering political support to more high-ranking politicians. This involved harassing members of rival factions, coercing villagers into voting for particular electoral candidates, and accumulating land and other wealth by force in order to consolidate their local strangleholds. It was only once they had significant wealth and political influence that politicians could afford to get others to do their dirty work for them. Moreover – as the relatives of Chowdri Mazhar Ali told me – higher level politicians were in the public eye to a greater extent than local ones and therefore needed to protect their reputation.

Thus the aggressive behaviour of Chowdri Abdullah and his siblings established them as effective local politicians and over the years local Jats increasingly approached them to get help with land and irrigation disputes and to intercede with the police on their behalf. A reputation for this sort of forcefulness was also an asset for the simple reason that it instilled fear in the majority of the inhabitants of the area who had neither the wealth, the connections nor the muscle-power to defend themselves. As the following chapters will illustrate a reputation for forcefulness could induce people to vote for one candidate over another because of fear of retribution if they voted the wrong way. Crucially, a reputation for forcefulness was a deterrent against predatory rivals on the watch for any sign of weakness.

And, last but not least, the use of force allowed politicians to forcefully acquire assets such as land and to engage in violent trades such as heroin trafficking. Chowdri Abdullah and his siblings made a great deal of money through the heroin trade and greatly increased their landholdings by occupying by force some of the best land that belonged to the Makhdooms, as

---

[13] On yet another occasion, during a wedding in Lahore, Chowdri Rafiq had driven off in the brand new car of an army colonel, whom he didn't even know. When the furious colonel came to the village to reclaim his car Chowdri Rafiq returned it, allegedly simply saying that he had taken the car for the fun of it.

well as significant areas of land belonging to smallholders in the area. They subsequently planted citrus orchards on much of this land, which also gave Chowdri Abdullah a handsome yearly income. In 2004 Chowdri Abdullah was said to have made a hundred lakh rupees from his citrus harvest.[14] His newfound wealth helped him increase his political influence and therefore gain clients whose votes he could deliver to more powerful regional politicians. Villagers expected that he would eventually contest in elections for the provincial assembly, although his rivals claimed that his reputation as a criminal would impede any such progress.

By 2006 Chowdri Abdullah had become the wealthiest and most influential chowdri in the village of Bek Sagrana. Given that politics was a zero sum game where one person's gain was another's loss, Chowdri Abdullah's upward mobility had set him on a collision course with members of the Ghulam Ali Ke faction. As Chowdri Abdullah grew wealthier he gained clients at Chowdri Haq Nawaz's and Chowdri Nawaz Ali's expense. The extent of this was such that by 2004 villagers were commenting that the Ghuam Ali Ke faction, whose patrons had been out of power since General Musharraf's military coup, was finished (*khatam*).

---

[14] In addition to taking significant amounts of land from the Makhdooms and others, he also purchased large plots of land. His purchase of 80 acres in the lowlands adjacent to the village of Bek Sagrana in 2004 was a particularly sore point for Chowdri Haq Nawaz and Chowdri Nawaz Ali, and it had almost resulted again in the eruption of armed conflict between the two factions. Chowdri Haq Nawaz and his close supporters worried that Chowdri Abdullah's purchase of the land would make him more powerful than he already was, and they were set on preventing the purchase from taking place. The owner of the land in question was a zamindar who lived in another village and belonged to another biraderi. He wished to sell the land in order to buy some closer to his home village. As soon as he learned that a deal was underway Chowdri Nawaz Ali, who was a friend of the seller, tried to get him to withdraw from his deal with Chowdri Abdullah and to sell it instead to a close relative of Chowdri Haq Nawaz. When the seller did try to withdraw, Chowdri Abdullah sent his gunmen to occupy the land and thereby prevent Chowdri Haq Nawaz's faction from appropriating it. In the end armed conflict did not occur because the seller was forced to sell the land to Chowdri Abdullah since the latter had already put a down payment on the land of 23 million rupees. Several months later Chowdri Abdullah proudly looked on as hired machinery began developing his newly purchased land. Villagers commented on how much Chowdri Abdullah's fortunes had soared, and rumours emerged that his new political allies in power, the Awans and the Melas, were planning to seek funds to build a minor distributor canal (*mogga*) that would irrigate Chowdri Abdullah's newly purchased plot.

By 2006, Chowdri Abdullah had purchased a petrol station[15] on the road between Sargodha and Lahore and he also ran the only bus service to Sargodha that passed by the village. He had also purchased property in Sargodha as well a new house in Lahore's Model Town. It even appeared that following a trip to Dubai he had decided to invest in real estate there. In 2005 Chowdri Abdullah caused a great stir by spending 80 million rupees on a Toyota Lexus Land Cruiser that he personally imported from Japan. Upon its arrival a large number of villagers assembled at his dera to watch Chowdri Abdullah proudly displaying his new vehicle. Additionally, while Chowdri Abdullah hadn't even completed secondary school, his eldest son was now attending the Lahore University of Management Studies (LUMS), one of the costliest and most prestigious higher education institutions in Pakistan.[16] On the political front Chowdri Abdullah aligned himself with politicians in the ruling PML-Q and became Union Council Nazim in General Musharraf's devolved government scheme.[17] Every day dozens of people from Bek Sagrana and surrounding villages gathered at his farm house to seek help with some problem or other. Others simply came to pay their respects and to thereby assure Chowdri Abdullah of their continued loyalty.

Chowdri Abdullah presided over them with absolute authority from a heavy gold painted sofa that looked like a throne. From there he attended and made phone calls and heard people's problems. He would occasionally interrupt the assembly by publicly performing his prayers or by vigorously performing his morning exercise in the courtyard where clients and servants could admire his physical strength. Other times he would interrupt the assembly in order to speed away to the local *thana* in his Land Cruiser.

For several reasons Chowdri Haq Nawaz proved unable to rise to the challenge posed by Chowdri Abdullah. One was that his patrons, Chowdri Mazhar Ali and Chowdri Nawaz Ali, were out of office: Chowdri Mazhar Ali had been put under house arrest for a year following General Musharraf's coup – because of his close ties with Nawaz Sharif – and Chowdri

---

[15] One of his employees at the petrol station told me that Chowdri Abdullah meticulously kept track of the accounts and called him no less than twice a day in order to do so.

[16] Chowdri Abdullah told me that when his son had completed his studies at LUMS he hoped to send him abroad to the London School of Economics to study law.

[17] He was succeeded in the post by his younger brother in 2006.

Nawaz Ali had left Nawaz Sharif's PML-N to join a splinter group of the PML-Q in order to escape his younger brother's fate.

On the other hand Chowdri Abdullah had managed to align himself and befriend MPAs in General Musharraf's ruling PML-Q. Another factor was that he belonged to a smaller and less aggressive sibling set than Chowdri Abdullah. Chowdri Haq Nawaz' youngest brother had died, and his elder sibling was known as a somewhat miserly recluse who from his mid-forties showed much more interest in mysticism than in politics. Thus although Chowdri Haq Nawaz had several supporters both in and out of the village, his village faction lacked the unity of purpose and core strength that the Lambar Ke faction enjoyed. What's more, most villagers, including close relatives who supported him, believed that Chowdri Haq Nawaz lacked the charisma, force, courage, and even the intelligence of Chowdri Abdullah and his siblings.

Despite villagers being highly critical of Chowdri Abdullah because of his occasional cruelty and criminal activities, most of them thought that he and his siblings had several important qualities that were not in equal evidence in Chowdri Haq Nawaz. As mentioned above, the forcefulness and energy of Chowdri Abdullah and his brothers meant that, whatever else, villagers believed them to be effective patrons. In addition most villagers believed that despite their hardness (*sakhti*) there was some evidence that they also had a compassionate and generous side to them.[18] While he lived Chowdri Rafiq had demanded absolute loyalty from servants and gunmen but in exchange he had often been lavishly generous. Two of his ex-gunmen told me that although he had been highly exacting and at times unpredictable, he had also rewarded loyalty and obedience through patronage and gifts of money. They told me that unlike other chowdris, including Chowdri Haq Nawaz, he never let any of them rot in jail when they had been caught by the police in a gunfight or while carrying out criminal activities on their master's behalf. Similarly, even though Chowdri Abdullah was also highly demanding of servants and clients, he too was known to be very generous. Villagers often illustrated this with the fact that he had gifted four acres of fertile canal-irrigated land to an ageing gunman who had served both him and his father before him with absolute loyalty. He was also known to be generous when he felt repentant for having been too harsh towards a client or a servant. For example, on one occasion he repented for having beaten and verbally abused a servant for a minor offence by making him a gift of

---

[18] People often told me that although they were hard on the outside they were soft on the inside (*bahar se sakht lekin andhar se naram*).

several thousand rupees and by even publicly asking him for forgiveness.[19] This side of his personality helped increase Chowdri Abdullah's stature, as did the fact that he had a sense of humour. When he wasn't angry about something, he often cracked jokes with his clients when they assembled at his dera.

On the other hand Chowdri Haq Nawaz was held to possess few such admired qualities. Like Chowdri Abdullah he also wished to be the most influential man in the area of Bek Sagrana. He too was involved in a variety of criminal activities, and like Chowdri Abdullah he was also reputed for being harsh and at times cruel. The similarities ended there, however. In the first place, Chowdri Haq Nawaz was not as brave as his rivals and many villagers went so far as to claim that he was a coward. One event that was often invoked to illustrate this related to an episode that took place years before I carried out my fieldwork. At that time, Chowdri Haq Nawaz had been in the business of organising cockfighting and gambling events in the village. Because gambling was illegal no such event took place unless the person organising it could assure participants that the police would not turn up and arrest them. The organiser generally did this by paying a low-ranking police officer to warn him if the police were intending to patrol the area. For arranging this sort of insurance for the participants the organiser of the event received 10 per cent of the proceeds from the gambling on the cockfights and a third of the profits from the dice games. The policeman involved would also receive a share of the profits from the cockfights and another third of the profits from the other gambling activities. During one such event the police turned up despite Chowdri Haq Nawaz having obtained assurances from his police contact that they would not. Instead of facing the police and protecting those attending the event from arrests and police brutality, Chowdri Haq Nawaz's immediate reaction had been to flee. As a result several villagers had been badly beaten and a few arrested by the police. The villagers who recounted this event claimed that if Chowdri Abdullah had been the one responsible for the event the police wouldn't have turned up in the first place but that, even if they had, Chowdri Abdullah would never have run away from them.

To the villagers who related this episode, it clearly illustrated that Chowdri Haq Nawaz was not the sort of person who could protect his clients from

---

[19] Asking for forgiveness was generally seen as somewhat demeaning, and it was generally only the defeated and the socially subordinate to whom it was thought appropriate. However, on this occasion Chowdri Abdullah showed that he was above pride and capable of the humility that was characteristic of God-fearing men.

the authorities, including the police. It was also representative of other failings that undermined the public perception of him. While some of his rivals, notably Chowdri Rafiq, fought alongside their gunmen, Chowdri Haq Nawaz preferred to stay behind and direct operations from a distance. Like his rivals Chowdri Haq Nawaz also demanded absolute loyalty and dedication from gunmen as well as other servants and clients, but unlike his rivals he often failed to reward those who served him with generosity and patronage. It was virtually a matter of routine for gunmen, who after all did much of the dirty work of the landlords they served, to find themselves in jail sometimes. When this happened, it was expected that their patron would do what was within his power to secure their release, and meantime that he would supply them with cigarettes and even food while they were there.[20] Although Chowdri Abdullah didn't always look after the welfare of his gunmen Chowdri Haq Nawaz's track record on this score was far worse. Many of Chowdri Haq Nawaz's gunmen complained that he did little to help them when they were in trouble.

The combination of the fact that he wasn't part of a large and cohesive sibling set, coupled with what was perceived to be a pleasure-seeking personality lacking in force, courage and generosity, made Chowdri Haq Nawaz a relatively unpopular and unsuccessful factional leader. Unlike his opponents Chowdri Haq Nawaz neither acquired new land nor set up any new business ventures. While by 2006 Chowdri Abdullah had acquired significant new landholdings and established various businesses, Chowdri Haq Nawaz possessed little more than what his father had passed on to him. This consisted of approximately 70 acres of land and a few buildings that he leased out for commercial activities in a nearby market town. People often noted that while Chowdri Abdullah cruised around in his brand new Lexus Land Cruiser, Chowdri Haq Nawaz continued to drive around in an old Toyota pickup truck, though even more commented upon was the fact that Chowdri Haq Nawaz's father continued to travel around in a battered old Hyundai that broke down regularly. In addition to these things, unlike Chowdri Abdullah, Chowdri Haq Nawaz was never able to obtain any form

---

[20] If a certain landlord managed to obtain the release of his men before his rival was able to secure the release of his, and if the former's were better cared for while in jail than the latter's were, this would serve as evidence that the latter either lacked influence or was careless, or perhaps even that he was miserly. A reputation for lack of influence and miserliness could badly damage the political influence of a landlord. To avoid this, rival landlords often competed to show off how efficiently they could get their gunmen out of jail, and how they supplied their gunmen with better food and cigarettes while they were incarcerated.

of political office. He rarely came to the village and even when he did few clients assembled at his farmhouse.

The fact that despite all of this Chowdri Haq Nawaz continued to have supporters is largely attributable to Chowdri Nawaz Ali and to Chowdri Mazhar Ali. Many of the supporters of the Ghulam Baksh Ke faction supported Chowdri Haq Nawaz because by doing so they also supported Chowdri Mazhar Ali and Chowdri Nawaz Ali who – despite being out of office – still had connections with people in authority and who were likely to return to power in the future. Chowdri Nawaz Ali's patrons were aware that he had become a liability to the faction. They complained that his lack of popularity cost them votes, and as will be shown in chapter five it even led them to seek another candidate from the Ghulam Baksh Ke faction of Bek Sagrana to contest the union council elections in 2005. However, because of their close kinship ties to him they continued to support Chowdri Haq Nawaz. Furthermore, even though they were aware of his shortcomings, they continued to support him because he was the person most committed to fighting the growing influence of Chowdri Abdullah, whom he regarded as a personal enemy.

Enmity and instrumental alliances

The conflict between the Ghulam Baksh Ke faction of Chowdri Haq Nawaz and the Lambar Ke faction of Chowdri Abdullah began to rage in the early 1990s, when Chowdri Abdullah's growing influence became a serious threat to Chowdri Haq Nawaz's local leadership and therefore, indirectly, to Chowdri Nawaz Ali's influence in Bek Sagrana. By this time the Makhdooms had long ceased to be an external force that united the Gondals against a common enemy. Moreover Chowdri Abdullah's growing influence meant that he no longer needed to rely on his Gondal relatives from Bek Sagrana.

At the time when the conflict began the leaders of both factions still lived in two large fortress-like houses adjacent to each other in the centre of the village. Villagers recalled a not so remote time when not a week passed without a shootout taking place between the houses of Chowdri Haq Nawaz and Chowdri Abdullah. Bullet marks were still visible on the walls of both houses in 2004 after the conflict had subsided and both Chowdri Haq Nawaz and Chowdri Asghar – Chowdri Abdullah's youngest brother – were rendered lame by bullet wounds suffered during these exchanges of fire. Villagers told me that during these shoot-outs at least three gunmen belonging to both parties were killed and several others imprisoned. The leaders of both factions and their close followers fought each other mainly over elections. These disputes were often complicated by the fact that both sides tried to entangle each other in cases, some real and some fabricated, with the police and the judiciary.[21] The causes of these weekly gun battles were often real and perceived personal slights by an opponent and his men. The fact that a gunman didn't like the way a rival gunman had looked at him was enough to result in an exchange of insults, which could in turn escalate into a full blown shootout where the landlords themselves became involved.[22]

These battles made life difficult not only for the landlords and their men but for villagers as a whole. Not only did villagers have to put up with the dangers of stray bullets, they also had to put up with gunmen who began abusing their power. Villagers told me that some of the younger gunmen had started behaving as if they were landlords by bossing people around, sending them off on errands, simply taking supplies from shops and stealing people's poultry and livestock. Far from being like the Robin Hood type figures described by Eric Hobsbawm in *Bandits* (2000), gunmen stole from the poor and were regarded as the stooges of the rich and as traitors to their class. Either out of indifference or because they were too busy fighting each other, neither Chowdri Haq Nawaz nor Chowdri Abdullah appear to have done anything to discipline their men.

Some of the wealthier landlords in the village, including Chowdri Abdullah himself, were able to relocate their compounds outside the village in an effort to gain some respite from the flare-ups in fighting. Chowdri Abdullah and his brothers built their new headquarters amidst citrus orchards about half a mile away from the village, where their family and dependants could spend their days in peace without having to fear stray bullets. Similarly Chowdri Haq Nawaz's reclusive elder brother moved to a peaceful location half a mile south of the village. These moves, however, did not signal the end of the conflict, which continued to rage for several years after and which subsided only following the death of Chowdri Rafiq Gondal. For the villagers themselves even worse times still lay ahead.

The leaders of both factions continued to fight each other both overtly and covertly through often intricately Machiavellian schemes. In order to combat the growing influence of Chowdri Abdullah and his fearless sibling Chowdri Rafiq, Chowdri Mazhar Ali decided to use his influence to help

---

[21] See chapter five for examples of this during the 2005 local body elections.
[22] According to the village mullah at least 90 per cent of the shootouts at the time had begun because of quarrels amongst gunmen. Although I do not necessarily accept this statistic at face value it does suggest that gunmen played an important role in instigating the fighting.

Chowdri Haq Nawaz. He did this by providing both logistical and political support to four young and notorious Gondal siblings who lived in Bek Sagrana. The four siblings in question each owned approximately 15 acres of land and had become involved in several criminal ventures including bootlegging, gambling and, to a lesser extent, drug dealing. Villagers told me that because their father had never encouraged them to study they had since adolescence directed their energies towards crime. Chowdri Mazhar Ali essentially offered the four siblings political support in exchange for them to unite with Chowdri Haq Nawaz and take up arms against Chowdri Abdullah. Chowdri Mazhar Ali even sent several of his men to purchase weapons for them from a contact in the Federally Administered Tribal Areas (FATA) bordering on Afghanistan.

Initially the four siblings fought fiercely against Chowdri Abdullah. In one of the various shootouts Chowdri Abdullah's youngest sibling, Chowdri Asghar, was shot in the leg causing him, like Chowdri Haq Nawaz, to limp for the rest of his life. Over time, however, the four siblings grew increasingly assertive and independent from both Chowdri Haq Nawaz and Chowdri Mazhar Ali. Since most of the influential Gondal landlords had by then moved out of the village the four brothers began what was effectively a reign of terror within it. On several occasions the four brothers sexually assaulted Mussalli women during the night. They also wandered around the village carrying weapons and bullied anyone who crossed their path. One hot summer day they came upon a carpenter (Tarkhan) who was walking towards them and who inadvertently failed to give way to them. Because of this the four brothers abused him verbally and then made him take off all of his clothes and lie down on the burning hot tarmac. Villagers told me that during this time even some of the wealthiest and most influential chowdris in the village feared the four siblings and avoided the village if they could.

As discontent amongst Gondals and other inhabitants of Bek Sagrana grew Chowdri Mazhar Ali and Chowdri Haq Nawaz tried to reign in the four siblings whom they had initially abetted, but failed to do so. In the end the alliance not only came to an end but the four siblings joined the Lambar Ke faction of Chowdri Abdullah against whose members they had only recently been fighting. However, this new alliance proved to be equally short-lived as the four siblings continued making life unpleasant for everyone in Bek Sagrana. As discontent amongst villagers grew Chowdri Abdullah and Chowdri Shahawaz were compelled to get together for the first time in years in order to do something about the four brothers. In the end the two leaders managed to organise a police raid which landed all four siblings in jail. After a couple of months the mother of the four young men went to beg Chowdri Abdullah for their release and her request was eventually granted on the condition that the four young men ceased to cause trouble.[23]

Chowdri Abdullah and his siblings also concocted schemes to weaken and possibly eliminate the leaders of the Ghulam Baksh Ke faction with the help of outsiders. Prior to General Musharraf's takeover in 1999 Chowdri Abdullah had become expediently allied with the influential politician Chowdri Ghulam Ali, who at the time was still in the middle of a longstanding feud with Chowdri Mazhar Ali and his brothers. Although Chowdri Ghulam Ali Malkana belonged to a branch of the extended Gondal biraderi known as the Malkanas,[24] the genealogical distance between him and Chowdri Abdullah was far greater than that between Chowdri Abdullah and Chowdri Mazhar Ali. Like Chowdri Mazhar Ali, Chowdri Ghulam Ali Malkana first became a member of the national assembly in 1985 under General Zia-ul Haq and subsequently shifted his allegiance to the PPP where he still remains.[25]

Chowdri Ghulam Ali Malkana's enmity with Chowdri Mazhar Ali began in the early 1990s and encompassed the rest of the period of civilian rule, which lasted from 1988 to 1999 and saw power transferred back and forth between Nawaz Sharif and Benazir Bhutto. Following the end of General Zia-ul Haq's regime Chowdri Ghulam Ali Malkana joined Benazir Bhutto's PPP while Chowdri Mazhar Ali joined Nawaz Sharif's PML-N. The feud between them allegedly began during an electoral

---

[23] Two of the four brothers died shortly after their release in any case, which seemed to diminish the potential trouble-making inclinations of the surviving two. Of the two who died, one died of a heroin overdose and the other in a firearms accident at a drinking party. This tragic end was welcomed by almost all of the villagers, who believed that God had punished them for their misdeeds.

[24] Malkana basically referred to the people from the town known as Mid or Mid Gondal.

[25] Chowdri Ghulam Ali Malkana ate, dressed, and behaved like a Jat cultivator and took pride in it. He signed documents with a thumbprint and attended sessions in the provincial assembly wearing the traditional Punjabi clothes. One of his principal sources of revenue was widely known to be from buffalo theft. He was known to have not only sent thieves and retainers to carry out nightly raids but to have personally accompanied them as well. He possessed several trucks which he used to carry the buffalos away to the Frontier Province, where their owners would be unable to trace them, and he was also widely rumoured to own a place where he could hide stolen buffalos in a sparsely populated area bordering the Jhelum River. His personal farm in *Mid* Gondal boasted several hundred buffalos that grazed on open pastures.

campaign when Chowdri Mazhar Ali Gondal and Chowdri Ghulam Ali Malkana were running for the same seat in the Punjab provincial legislature. Violent feuds between families of politicians frequently began during elections when candidates tried to entangle their opponents in court cases prior to election day and to take over polling stations on election day itself.[26] In this case what appears to have sparked a feud – which lasted almost a decade and wound up costing several lives – was an otherwise incidental taunt. At the time 17-year-old Asadullah, Chowdri Mazhar Ali's youngest half-brother whose reckless bravado rivalled that of Chowdri Rafiq, was campaigning on behalf of his elder brother. At one stage during the campaign some of Chowdri Ghulam Ali Malkana's men encountered Asadullah and began to taunt him, telling him that his elder half-brother (Chowdri Mazhar Ali) was about to lose the upcoming elections. The taunt led to an argument and eventually culminated in the reputedly hot-headed Asadullah shooting and killing three of Ghulam Ali Malkana's supporters. Chowdri Ghulam Ali Malkana started a court case against Asadullah over the incident, which eventually led to further fighting between his men and Asadullah. Two more of Malkana's men would be shot and killed in the course of it.

Thanks in part to his elder brother's protection Chowdri Asadullah was able to spend almost seven years as a fugitive before he was caught and sentenced to death. Chowdri Asadullah's nephews, who seem to have been fond of Western movies, told me that throughout this period Asadullah had lived like a 'desperado', spending his time in hideouts with his faithful retainers. In the meantime Chowdri Ghulam Ali Malkana and his men had been looking for opportunities to take revenge (*badla*), meaning that for about 15 years Chowdri Nawaz Ali and Chowdri Mazhar Ali rarely travelled without a contingent of at least three gunmen.

Because Chowdri Abdullah and Chowdri Ghulam Ali Malkana shared their enmity towards Chowdri Mazhar Ali, they became allies. Thus Chowdri Abdullah provided Ghulam Ali Malkana with votes when he contested elections for the provincial assembly prior to Musharraf's military coup. Chowdri Abdullah's motivation was simply to prevent Chowdri Mazhar Ali from winning those elections, not to help Ghulam Ali Malkana consolidate his power. In fact according to many people it was actually against Chowdri Abdullah's interests for Chowdri Ghulam Ali Malkana to become too powerful: people said that his ultimate goal was to eliminate both Ghulam Ali Malkana and Chowdri Mazhar Ali from the political scene in order to emerge as the most powerful politician in the area himself.

---

[26] See chapter five.

To this effect Chowdri Mazhar Ali claimed that in 1998 Chowdri Abdullah had managed to make him unwittingly grant protection to a group of wanted men who planned to launch an attack on Chowdri Ghulam Ali Malkana and murder him. Chowdri Abdullah's basic plan was to eliminate Ghulam Ali Malkana – despite his ostensible political alliance with him – and to entangle Chowdri Mazhar Ali in a court case for having harboured the men who killed him. In the event, Chowdri Abdullah managed to get a group of wanted men – who had their own grudge against Malkana – to ask one of Chowdri Mazhar Ali's clients for protection from the police. Chowdri Abdullah knew that this client would then ask Chowdri Mazhar Ali to provide them with a place to hide.[27] Chowdri Mazhar Ali complied with his client's request and allowed the wanted men to hide at his dera where it was assumed, because he was still a provincial minister at the time, the police would not intrude. In the end, however, Chowdri Abdullah's plan failed when the police did in fact raid Chowdri Mazhar Ali's dera and arrested the men before they could launch their attack on Chowdri Ghulam Ali Malkana. Nevertheless, Chowdri Abdullah appears to have been partially successful because the incident caused Chowdri Mazhar Ali to become temporarily embroiled in a police case. Eventually the alliance between Chowdri Abdullah and Chowdri Ghulam Ali Malkana ended because Chowdri Mazhar Ali established a truce with Ghulam Ali Malkana by agreeing to support him in future elections and by paying him blood money (*diyat*) for the murders of his supporters by Chowdri Asadullah.[28]

It would take the military takeover by General Musharraf some months later to achieve, at least in part, what Chowdri Abdullah had been unable to accomplish through his own devices. Chowdri Mazhar Ali became entangled in a variety of NAB cases on several corruption charges (see chapter two) and remained out of office for the nine years that General Musharraf was in power. This situation provided Chowdri Abdullah with an invaluable opportunity to strengthen his own political clout further at the expense of Chowdri Mazhar Ali and Chowdri Haq Nawaz. Shortly before that happened, however, Chowdri Abdullah's brother Rafiq lost his

---

[27] The individual in question was a friend of both Chowdri Mazhar Ali and Chowdri Abdullah. He was not as powerful as Chowdri Mazhar Ali and could not have personally granted the wanted men any significant form of protection.

[28] Under this settlement Chowdri Ghulam Ali Midhiana agreed to receive 25 acres of land from Ghulam Ali as compensation for the murders. Although many details of the agreement remained secret it appears to have been facilitated by another political figure in Sargodha who had an interest in uniting the two parties in order to defeat Khuda Baksh Mekan in upcoming elections.

life. Several Gondals believed that the encounter with the police which caused his death had been orchestrated by Chowdri Mazhar Ali, who was particularly keen to get rid of the intrepid young man and thereby weaken the Lambar Ke faction.

The political realignments that resulted from General Musharraf's takeover provided Chowdri Abdullah with an opportunity to move out of the opposition and to join forces with the newly formed ruling coalition, and he aligned himself with Chowdri Khuda Baksh Mekan who was now an opponent of Chowdri Mazhar Ali. Chowdri Abdullah also aligned himself with Chowdri Zahid Iqbal Awan, who had also joined the ruling coalition and who was a long-standing enemy of Chowdri Mazhar Ali. When young, the latter and his servants had killed Zahid Iqbal Awan's younger brother in a college brawl.[29] However thanks to his contacts Chowdri Mazhar Ali managed to spend only one year in jail – although one of his servants was less lucky and spent decades in jail awaiting a death sentence. Since then Chowdri Awan had become Chowdri Mazhar Ali's most powerful opponent in the district.

Thus Chowdri Abdullah once again aligned himself with the enemy of his enemy and with non-kin against a kinsman. Chowdri Abdullah now found himself aligned against the newly reconciled forces of Chowdri Ghulam Ali Malkana and Chowdri Mazhar Ali who had joined forces against the ruling PML-Q. They agreed to support each other in the local council elections of 2005 as well as in upcoming provincial and national assembly elections. At one stage there were even rumours that the alliance between the two would be consolidated through a marriage between Chowdri Mazhar Ali's son and Chowdri Ghulam Ali Malkana's daughter. In this manner two longstanding enemies affiliated with traditionally opposed political parties (the PPP and the PML-N) joined forces against the PML-Q, thereby reducing the number of local political groupings from three to two.[30]

---

[29] Chowdri Nawaz Ali only spent one year in jail after the murder and before launching on his legal and political career. He and his relatives claimed he got such a light sentence thanks to his father's contacts in the judiciary and the police. An alternative explanation offered by villagers suggests that his sentence was light because his servant, who had been present during the attack, bore the full weight of punishment in his place. The servant, whose father was Muhammad Ali's tenant, languished in jail for his entire life.

[30] My informants told me that it was common for politicians to promise marriage alliances just before elections, but that these promises were frequently broken after elections when one party or the other had obtained what it wanted from the other.

The forging of the new political alliance between Chowdri Ghulam Ali Malkana and Chowdri Mazhar Ali fit the model noted above of a temporary coincidence of interests resulting from shared opposition to a third party. Despite the fact that the leaders of the two factions began to assert that they were brothers by virtue of belonging to the same extended Gondal lineage, their long history of enmity and conflict meant that they continued to distrust each other intensely. Chowdri Asadullah, who had ignited the feud between his half-brother Chowdri Mazhar Ali and Chowdri Malkana, couldn't fully reconcile himself to the alliance and his elder siblings tried to keep him away from the negotiations with Malkana in case old tensions were revived. For their parts Chowdri Mazhar Ali and Chowdri Nawaz Ali didn't fully trust Chowdri Ghulam Ali Malkana either, whom they believed still harboured grudges against them. More than that perhaps, they were aware that the coincidence of interests that emerged from General Musharraf's military takeover might not last forever. Since both Chowdri Nawaz Ali and Chowdri Ghulam Ali Malkana remained loyal to their respective political parties they recognised that they might become opponents once Musharraf and the PML-Q were out of office and the PPP and the PML-N resumed hostilities in their bid to gain political power. A more immediate obstacle to their alliance related to whether they would be able to agree on what offices Chowdri Nawaz Ali and Chowdri Ghulam Ali Malkana would run for. Although they cooperated during the local council elections of 2005, they feared that tensions might arise over who would run for the local provincial assembly seat in the upcoming elections. The concern arose from the fact that Musharraf's government ruled that members of legislative assemblies must hold bachelor's degrees. As a result they had agreed that Chowdri Nawaz Ali, who had a degree in law, would run for the provincial assembly seat while Chowdri Ghulam Ali Malkana, who was illiterate, would run for the seat of tehsil nazim for which no degree was required. However, it was expected that these educational requirements might soon be scrapped.[31] This would present a problem for the alliance because it might cause Chowdri Ghulam Ali Malkana to want to run for the seat in the provincial legislature, creating a clash over it with Chowdri Nawaz Ali.

---

[31] The rule that only people with bachelor's degrees could run in elections for the legislative assemblies was widely held to have been designed by Musharraf's government in order to selectively prevent certain candidates from running. Members of the opposition demanded that the rule be scrapped as it excluded the vast majority of Pakistanis.

Conclusion

This chapter showed how enmity could be a more powerful force in the creation of alliances between politicians than extended kinship ties in the form of the biraderi. What had historically held the Gondals of Bek Sagrana together was their opposition to the politically dominant Makhdooms rather than corporate unity arising from shared ties of descent. So long as the Makhdooms had been politically dominant it had been in the interest of the Gondals of Bek Sagrana to set aside their differences in order to overtake them politically. However, following the decline of the Makhdooms fault lines between the Gondals gradually became more distinct until eventually, in the early 1990s, internecine conflict erupted between the leading Gondal households of the two village factions. As in Barth's analysis of the Swat Pathans the fault lines that emerged within the Gondal biraderi were the result of agnatic rivalry arising from competition for local dominance.

However because control over the state was central to the reproduction of landed power, it was competition to control it through electoral politics – rather than simply to control the land – that gave rise to factionalism among agnates. Moreover factional conflict in rural Punjab wasn't always necessarily between agnates, and different villages were divided on other bases. While in villages dominated by a single biraderi, as in Bek Sagrana, conflict might take place between agnates with varying degrees of genealogical proximity between them, in other villages with two zamindar biraderis conflict might predominantly follow biraderi-based cleavages. Other villages might even possess more than two factions.

Having demonstrated how competition for political and economic supremacy in the village fuelled local rivalries, the second section of this chapter sought to illustrate the role of enmity between village factional leaders in determining their broader political alliances. It showed how factional leaders almost invariably became the allies of the enemies of their enemies. Because the only thing that held many of these alliances together was a temporary sharing of interests based on a shared enmity they were often short-lived. For example, the alliance between Chowdri Abdullah and Chowdri Ghulam Ali Malkana was instrumentally based on their joint enmity with Chowdri Mazhar Ali and not on a desire to see each other succeed; indeed, as was noted, according to many people Chowdri Abdullah secretly harboured a desire for and plotted the demise of Chowdri Ghulam Ali Malkana. The temporarily instrumental nature of the alliance was further demonstrated when it fell apart shortly after General Musharraf took over in 1999. The alliance that then emerged between Chowdri Mazhar Ali and Chowdri Ghulam Ali Malkana was equally provisional as there was little love lost between both parties after a long history of enmity that had cost money and land as well as several lives. In this case their alliance was based on a common desire to counter the new government. Once that government fell, however, there was a distinct possibility that the alliance might fall apart.[32]

The highly instrumental nature of many political alliances raises the question of exactly what role kinship played in their creation in rural Punjabi politics. The chapter demonstrated that the extended biraderi was generally not the basis for political decision-making among powerful landlord politicians. Instead close kinship ties, including those between male siblings and those between close relatives united by marriage ties, were often at the core of political factions. The political significance of cooperation between male siblings was illustrated by the case of Chowdri Abdullah and his brothers.[33] The fact that Chowdri Abdullah belonged to a set of four forceful brothers played a decisive role in the political success of his faction. This gave him an advantage over Chowdri Haq Nawaz who, following the death of his younger brother, was left with only his older, who preferred to keep his distance from politics. Nevertheless, Chowdri Haq Nawaz's close and variously intertwined kinship relations with Chowdri Nawaz Ali and his siblings permitted him to retain his influence in the village.[34]

Following Hamza Alavi (1972a) the material presented can also be taken to indicate that the political role of the extended biraderi varied according to the socio-economic rank of its members. Alavi argued that the least cohesive biraderis were those that were economically and politically dependent upon the landlords. Their dependency meant that they were unable to create horizontal ties of solidarity amongst themselves. Similarly he argued that biraderi solidarity tended to be weak amongst big landlords. The reason for this was that such landlords didn't need to rely on their extended biraderi for political support but could instead rely upon their clients, retainers and economic dependants for it. In contrast, the most

---

[32] The only thing that might have helped hold the political alliance together would have been the additional establishment of a marriage alliance.

[33] Likewise Pettigrew (1975) shows how in East Punjab strong ties between siblings were crucial to political success.

[34] This having been said, it should be noted that there were cases in which brothers could be political rivals. One famous instance involved two Gondal politicians from the area who, because of a dispute over their inheritance, had become fierce political rivals. However, cases such as these appear to have been uncommon and were met with strong disapproval.

cohesive biraderis were those of economically independent landholders.[35] Their economic independence meant that they could unite and assert their interests against those of the bigger landlords. These biraderis often voted as a block, and powerful landlord-politicians spent a great deal of time trying to gain their votes during elections. In accordance with this analysis it is arguably the case that the Gondal biraderi of Bek Sagrana had been more cohesive before the political and economic fortunes of certain of its individual members began to soar. Earlier the Gondals of Bek Sagrana had depended upon each other in order to achieve political influence and combat the Makhdooms. This changed when people like Chowdri Nawaz Ali, Chowdri Mazhar Ali and Chowdri Abdullah acquired greater power and wealth, allowing them to consolidate their personal client bases and depend less upon other Bek Sagrana Gondals to assert their political power. In Bek Sagrana it also translated into an erosion of endogamous ties. As Chowdri Abdullah gained power and his interests started to clash with those of his rivals in the Ghulam Baksh Ke faction, his siblings and children started establishing marital alliances outside the local Gondal, and Ranjha, biraderis in the district. Thus Chowdri Abdullah's two younger siblings both married into a branch of the Sial clan from the neighbouring district of Jhang rather than into a local Jat clan. Although Chowdri Mazhar Ali talked about marrying his second son to Chowdri Abdullah's daughter in order to end hostilities this never happened because both sides felt that their relationship was beyond repair. In the end there was talk of Chowdri Mazhar Ali's son marrying into a Lahori industrialist family, and of Chowdri Abdullah's daughter also marrying into family from Hafizabad that was living in Lahore.

The findings of this chapter are important for understanding the political sociology of the Pakistani Punjab because they shed light upon some of the reasons why so few candidates in the central and northern Punjab win decisive victories even when their biraderi is the numerically dominant one in the area. As Wilder puts it, '... because nearly every village is divided into factions, and these are often linked into larger factional groupings, their cumulative effect can be great. This poses a challenge to political parties seeking to increase their support in rural constituencies. No matter which candidate is selected, strong opposition is virtually guaranteed by rival factions' (Wilder 1999: 177).

The material I presented here confirms Ayesha Jalal's claim that military interventions have made rural politics more parochial and kinship based,

---

[35] Hardiman (1982: 206) similarly argues that 'lesser' Patidars in Gujarat often acted as a class and didn't necessarily follow the dictates of 'superior' Patidars.

but the chapter also expanded upon this insight. It showed how immediate siblings rather than extended biraderis tended to command people's political loyalties. I also suggested that the localisation of politics arguably brought personal enmity to the forefront of politics. Finally the chapter illustrated how the victims of feuds between landlords were most commonly ordinary villagers – factional leaders also suffered because of it, but they could escape it by sending their gunmen to fight in their place and by moving out of the village to newly built houses in their fields or to rented accommodation in town.

5
ELECTIONS AND DEVOLUTION

Devolution

This chapter examines whether elections empowered people and made their leaders more accountable. While various studies and reports – including those by authors such as Wilder (1999, 2004) and Waseem (1994) and by the ICG (2004b, 2005b) – emphasise how national-level electoral manipulations undermine democracy, few pay attention to how local power structures do so. These studies provide valuable evidence of the various machinations to retain power employed by authoritarian central governments in Islamabad. These include various forms of electoral and pre-electoral rigging as well as the holding of non-party elections, devolution programmes and national referendums. Wilder has claimed that a consistent theme running almost throughout the seven national party–based elections held in Pakistan prior to 2008 – as well as the various non-party and local party elections and presidential referendums – has been 'to legitimize the retention of power by the unelected institutions of the state rather than to transfer power to local institutions' (Wilder 2004: 102). This chapter focuses on how local power structures also undermine the electoral process and democracy. It examines how the Gondals were able to use their influence to undermine General Musharraf's attempt to allegedly empower the masses through a devolved government programme.

On 23 March 2000, General Pervez Musharraf declared that devolution 'was the beginning of a constructive, democratic, dynamic revolution whose sole objective is to place in [the] hands of the people the power to shape their own destiny . . . an unprecedented transfer of power will take place from the elites to the vast majority' (ICG 2004b: 5). The programme, devised and funded by various international organisations, aimed to encourage grassroots political participation and greater government accountability through the creation of various tiers of local government. According to the analysis of the World Bank (1998) in a report entitled 'A

Framework for Civil Service Reform in Pakistan' governance in Pakistan suffered from (a) an over-centralised organisational structure, (b) a serious lack of accountability to the public resulting from both a colonial legacy and years of military dictatorship and (c) the politicisation of civil service decision making and the failure of politicians to exercise their oversight role in the wider public interest.

LaPorte (2004: 155) similarly argues that the lack of accountability of Pakistan's civil service can be traced back to the early days of independence when Mohammad Ali Jinnah perpetuated the colonial viceregal tradition by designating himself governor general with the power to bypass ministers and deal directly with civil servants. After Jinnah's death, successive governors general largely retained these viceregal powers. Up to the time of Musharraf's takeover, the local embodiment of this centralised, top-down, bureaucratic power structure had been the deputy commissioner, who served as a one-man locus of judicial, executive and revenue functions at the divisional level. The deputy commissioner was largely an instrument of centralised government control who kept opposition politicians in check and rewarded political supporters. One fairly obvious consequence of this was that the concentration of powers in the deputy commissioner's hands created a vast potential for corruption and the abuse of power.

These and other analyses focus almost exclusively on the institutional factors that have impeded democratisation in Pakistan and are therefore incomplete. The district government – headed by the deputy commissioner – is subject to a variety of influences and doesn't somehow stand above the society it is meant to govern. As Matthew Nelson (2011) clearly illustrates, the district administration doesn't work in the top-down manner implied above but is in fact responsive to certain social pressures. He shows how village land registry officers (*patwaris*), who are formally controlled by the deputy commissioner, are in practice heavily influenced by landowners and members of the political establishment who make them tamper with land records for their private benefit and that of their supporters. Patwaris supplement their meagre salaries with bribes and favours from landlords and politicians in exchange for, among other things, delaying and cancelling mutations, misrepresenting the value of property held by insolvent owners and failing to enter the names of specific heirs. According to Nelson this means that the lack of democratic accountability in rural Pakistan is due to the landed elite's ability to systematically subvert the law rather than to a colonial legacy of bureaucratic authoritarian rule – as Jalal (1995) and LaPorte (2004) argue. Most frequently it is the landed elite's disproportionate power that deprives ordinary people of their legal rights – including the right to property – not an authoritarian bureaucratic structure or arbitrary authoritarian rule at the distant centre.

However, the Musharraf government ignored the issue of landed power and focussed solely on administrative reform. It wanted to give greater administrative powers to various elected tiers of local government and to abolish the powerful position of deputy commissioner. While this offered democratisation without pain – since it didn't challenge entrenched economic and political interests – it also offered the regime a chance to regain international legitimacy and access to foreign aid after two decades of political isolation. The implementation of a devolution programme was a good way to appease international calls for democratisation and to get foreign aid flowing once again. In the context of 'the war on terror' following the events of 11 September 2001, General Musharraf's devolution programme was presented to the international community as a way of increasing political stability in Pakistan and therefore, implicitly, as a way of avoiding the possibility of extremists taking control of the country and its nuclear assets.

While the Musharraf regime may genuinely have sought to increase 'grassroots' political participation, it had no interest in fostering democratisation at the provincial and national levels. There are good reasons to believe that General Musharraf was following in the footsteps of his military predecessors who through similar – though supposedly less radical devolution—schemes sought to institute lower tiers of government as a substitute for democratisation at the provincial and national levels. The ICG report of 2004b suggests that both General Ayub Khan's and General Zia-ul Haq's military governments used local government programmes in order to: '(1) depoliticise governance; (2) create a new political elite to challenge and undermine the political opposition; (3) demonstrate the democratic credentials of a regime to domestic and external audiences; and (4) undermine federalism by circumventing constitutional provisions for provincial political, administrative, and fiscal autonomy' (ICG 2004b: 1). General Ayub Khan introduced his Basic Democracy plan as a nominal concession to democracy after having suspended the constitution. He argued that the nature of Pakistani society meant that it wasn't ready for fully fledged democracy and that it needed the benevolent, modernising guidance of an enlightened elite. Under Basic Democracy, Pakistan was divided into 80,000 wards to elect a 'Basic Democrat' on a non-party basis, and local councils were created at the district as well as at the union, *tehsil* and *thana* levels. Roughly half of the members of the local councils were officially nominated, while those who were directly elected largely remained under the control of the district administration (which unlike under Musharraf's local government system remained unreformed, and retained the power to overrule council decisions and suspend the execution of their orders). General Ayub Khan also circumscribed provincial powers through the federal appointment of provincial governors. This meant that Ayub Khan's regime was able to create a new class of politicians over whom the central government could directly extend its control through the district administration. Many of these Basic Democrats were selected from gentry and middle-ranking landlord backgrounds in order to sideline powerful feudal landlords and to bypass both political parties and the provinces.

Additionally the Basic Democrats were used as an Electoral College for the presidency and the provincial and national assemblies. Because this Electoral College could easily be manipulated by the district administration (which was ultimately controlled by centrally appointed governors), it was used for the purpose of rubber-stamping the continuation of the Ayub regime. Thus, through the Basic Democrats General Ayub Khan was able to obtain a 95.6 per cent approval rating in a presidential referendum to extend his presidency for five more years. Thirty years later General Musharraf obtained a similarly high approval rating of 97.5 per cent in a presidential referendum thanks to his system of devolved government.

Following the rise of Zulfikar Ali Bhutto, Ayub Khan's Basic Democracy scheme was scrapped and it was almost another decade, following General Zia-ul Haq's military coup, before a similar scheme was implemented again. In its details General Zia's local government system differed somewhat from the Basic Democracy system, but its fundamental aim was the same: 'to cloak a highly centralised, authoritarian system of government under the garb of decentralisation' (ICG 2004b: 4). Like General Ayub Khan's system it aimed to establish an elite of faithful local politicians that could be manipulated through the district administration. General Zia used these local politicians to undermine the political opposition constituted by the popular PPP. He did this principally by staging non-partisan elections and using local governments as a means of extending patronage to pro-military politicians in order for them to win elections. Where his system differed from General Ayub Khan's was that local politicians didn't form the Electoral College for the national and provincial assemblies. Nevertheless, because the bulk of the finances of local governments came from federal transfers, this gave the central government ample scope to provide patronage to pro-military politicians to help them win elections. Following the restoration of democracy after General Zia's decade in power, his local government scheme was allowed to decay and was eventually scrapped in the early 1990s. It wasn't until General Musharraf's military coup in 1999 that the idea of local government returned to the national political agenda.

Although the planned reforms under General Musharraf's local government system were far more significant than the local government reforms carried out by both General Ayub Khan and General Zia-ul Haq, the evidence suggests that once again the scheme served to provide the garb of democracy and decentralisation to a highly centralised and authoritarian regime. Whereas neither of the previous two local government programmes had aimed to reform the district administration, General Musharraf proposed to do so radically by replacing provincially and centrally appointed bureaucrats with democratically elected local representatives. Under Musharraf's programme the indirectly elected district administration headed by the district *nazim* was, among other things, to take over the responsibility for law and order, education, health, agriculture, and transport from the traditional district management group and the deputy commissioner. Below the district level the tehsil administration was to take charge of municipal infrastructure, water, sanitation, and roads. Finally at union council level, the lowest administrative tier in the scheme, the union council administration was put in charge of registering births and deaths and also of encouraging the formation of voluntary associations termed Citizen Community Boards (CCBs) composed of a minimum of 25 unelected members. These voluntary associations were planned to encourage grassroots participation for the sake of monitoring service delivery and promoting accountability, as well as for the implementation of minor infrastructure projects. Additionally the union council was to serve as the Electoral College for the district nazim.

Despite these seemingly significant transfers of power to elected representatives, the evidence suggests that like previous local government programmes implemented by military rulers, Musharraf's devolution programme served to ensure regime survival by actually centralising power and fragmenting the political opposition. By forcing the provinces transfer 40 per cent of their total revenue to local bodies, the central government was able to take power away from provincial governments. The latter were disempowered to such an extent that many provincial ministers chose to swap provincial office for a district nazim post. The ICG even reports that because General Musharraf also bypassed members of the national assembly and dealt directly with district nazims, several of the former also swapped their posts for that of district nazim. The government also sought to bypass political parties by making the elections non-party based and to thereby prevent the emergence of large-scale opposition by the PML-N and by the PPP. The policy was allegedly meant to prevent the traditional political elite from capturing the local government system and to create a space for the emergence of a new, popular political class. According to the NAB chairman Daniyal Aziz it was also meant to prevent the politicisation of service delivery in the local government system. In the end it neither prevented the traditional elite from capturing the local government system nor the politicisation of service delivery.

By bypassing provincial governments the Musharraf's government, like the governments of his military predecessors, was able to create a political constituency that could be easily manipulated through the granting and withholding of government funds and patronage. This was reinforced in 2005 when it amended the local government ordinance to empower chief ministers to suspend the orders and decisions of local government bodies. A parallel amendment also made it easier for the chief minister to suspend and even remove district, tehsil and union council nazims.

Having the nazims under control, the government used them to mobilise electoral support by distributing funds and by politicising service delivery – which would according to the government have happened if political parties had participated in local body elections. Just as General Ayub Khan and General Zia-ul Haq had done before him, General Musharraf used local councillors in order to extend his term in power through a presidential referendum. During the April 2002 presidential referendum on extending Musharraf's term in office by five years it was widely reported, including in the village of Bek Sagrana, that the central government channelled funds through local governments to union councillors in order for them to campaign on Musharraf's behalf. Union councillors were said to have used the money to buy votes through cash payments as well as through minor infrastructure projects. Additionally, the funds were used to hire large numbers of buses in order to get supporters who didn't have means to travel to the polling stations.

In Bek Sagrana Chowdri Abdullah Gondal who had been elected union council nazim gave voters the additional incentive of a free meal at the polling station, allegedly by using government funds. In addition, the fact that in Musharraf's devolution programme the councillors served as a restricted Electoral College for the district nazim was further evidence of the way in which the devolution programme actually served the objective of entrenching Musharraf's regime rather than its stated objective of encouraging grassroots political participation and democratisation. Because union councillors could be easily manipulated through the granting and withholding of patronage it was easier to get them to vote for a pro-government district nazim than it was to influence an average of around a million voters per district. Once a pro-government district nazim was in place the government could then keep patronage flowing to its supporters in order to consolidate its power.

To further ensure favourable results in the presidential referendum – as well as in other local, provincial and national elections – General Musharraf's government is also widely reported to have taken measures that included harassing and even abducting political opponents and arbitrarily disqualifying opposition politicians from running for office. Politicians aligned with opposition political parties were barred from entering elections in both of the local body elections of 2001 and of 2005 on the grounds that these were supposed to be non-partisan, but politicians who were openly entering elections on a PML-Q platform were allowed to do so. Additionally, many candidates were disqualified for 'unspecified defects' in their educational qualifications, and others were disqualified for their alleged inability to prove 'adequate knowledge of Islamic injunctions'. Under the Local Government Ordinance of 2001 contestants for the positions of nazim and vice nazim (*naib-nazim*) were required to have no less than a secondary school certificate or matriculation for the alleged reason that this would raise the quality of politicians. In practice, however, this measure not only served to disqualify a large number of opposition politicians, including the PPP politician Ghulam Ali Malkana mentioned in previous chapters, but also to exclude the large majority of the population, who fell short of these standards. This contradicted General Musharraf's claim that the devolution programme aimed to 'transfer power from the elites to the vast majority'.

Another significant measure taken to ensure electoral success through the local government system was the large-scale manipulation of the judiciary through the transfer and replacement of judges by more pliable ones. This began at the highest level, that of the Supreme Court. This practice allowed Musharraf's government to place partisan session and district Judges, who since 1988 had acted as returning officers, at polling stations in order for them to either overlook or participate in various forms of electoral rigging. It also allowed the government to try either to ensure that polling stations were located in friendly territory or, if this wasn't possible, to locate polling stations in remote areas in order to make it difficult and costly for people to present themselves to vote. A large number of transfers were also widely reported to have taken place prior to both local body elections in various branches of the civil bureaucracy and the police. Yet other measures to ensure favourable election results included the large-scale gerrymandering of constituencies in order to divide the opposition. According to a report by the ICG, this was particularly aimed at the PPP in its Sindhi strongholds (ICG 2005b: 5).

The combination of all of these measures led to overwhelming victories for pro-government candidates in both the local council elections of 2001 and of 2005 and the presidential referendum of April 2001, when the proposal to extend General Musharraf's term in office by a further five years was approved by 97.5 per cent. This almost-farcically high approval in the referendum clearly pointed to widespread pre-election and election day rigging and led Musharraf himself to admit reluctantly that 'unbeknownst to him some of his supporters had shown over-enthusiasm' (Wilder 2004: 107). Nevertheless, following the 2005 local body elections and in situations where similar strategies were used by the central government, Prime

Minister Shaukat Aziz declared that the overwhelming pro-government turnout indicated that people had shown their support for President Musharraf and for his policies of enlightened moderation. However, as the following will show, neither democracy nor enlightened moderation was foremost on people's minds during these elections.

Devolution in Bek Sagrana

Bek Sagrana was one of the 13 villages that comprised the union council of Daulatpur. Unsurprisingly Daulatpur's first union council nazim elected in 2001 hailed from the *zamindar* class and was none other than the strongman Chowdri Abdullah Gondal. Two factors had greatly facilitated his becoming union councillor. The first was that his most powerful rivals, Chowdri Nawaz Ali and siblings, had been jailed and forced underground by Musharraf's government. This had cleared the way for him to grab the seat without any substantial opposition. The second was that he had obtained the backing of Chowdri Khuda Baksh Mekan, a powerful member of General Musharraf's ruling PML-Q who had supported his campaign by channelling government funds to him and directing the local police and civil service in his favour. In exchange Chowdri Abdullah had supported Chowdri Khuda Baksh Mekan's choice of district nazim and would deliver votes to Chowdri Khuda Baksh Mekan in future elections. This entitled Chowdri Abdullah to substantial patronage and gave him far greater influence than the position of union council nazim gave him formally.

Throughout my time in Bek Sagrana between January 2004 and July 2006 it became clear to me that neither Chowdri Abdullah, his political opponents, nor the villagers that the programme allegedly aimed to empower appeared to be particularly concerned with, or even aware of, the devolution programme's stated aim of democratisation and popular empowerment. Both Chowdri Abdullah and his opponents were principally concerned with the consolidation of their own power at each other's expense. Neither was concerned with even attempting to pay lip service to the idea that devolution was about creating greater government accountability through the empowerment of the masses. Chowdri Nawaz Ali and Chowdri Mazhar Ali largely saw the devolution programme as an attempt by Musharraf's government to extend central government control over politicians and to weaken the political opposition. Chowdri Nawaz Ali, who was an experienced politician and who had seen a similar programme implemented by General Zia, thought that the exercise was all a big show (*namuna*) and that it wouldn't survive beyond Musharraf's regime. As he was in the opposition he realised that for his and other allied factions to obtain the majority of union council seats in the district would be difficult.

He was keenly aware that the government would not only be dispensing funds to pro-government candidates but that the entire judiciary and civil services would be manipulated in order to support them. His hope was to at least prevent his rivals from entirely capturing the state.

In order to achieve this Chowdri Nawaz Ali joined forces with his erstwhile enemy Chowdri Ghulam Ali Malkana who, as a member of the PPP, was also in the opposition. As a part of the peace deal (*sula*) between them, which was agreed to only two months prior to the union council elections and followed years of violent feuding, Chowdri Nawaz Ali and Chowdri Ghulam Ali Malkana agreed to support each other in order to obtain as many of the 22 union council nazim seats in the tehsil of Qot Momin as possible. It had also been agreed that during the subsequent district nazim elections Chowdri Nawaz Ali's faction would get all of his allied union councillors to vote for a district nazim of Malkana's choice. Chowdri Nawaz Ali would also give all of his votes to Chowdri Ghulam Ali Malkana's candidate for the tehsil nazim elections. In exchange for this Chowdri Ghulam Ali Malkana would later support Chowdri Nawaz Ali in future elections for the provincial or national assembly. There were even rumours that the two parties would cement these deals with a marriage alliance in the near future.

The absence of popular empowerment and of public accountability on the part of politicians following the implementation of devolution is clearly demonstrated by the case of Chowdri Abdullah. To begin with, a large number of the 21 elected union council members whose role was to approve the budget and oversee the delivery of different services delegated all decision making to Chowdri Abdullah. If Chowdri Abdullah required their signatures for a given project, the union councillors gave it without any questions and I heard of no cases where they opposed his decisions. Part of the reason for this was that he had personally selected a number of his union councillors from among faithful subordinate clients and employees. One councillor, for example, was his sharecropper and another was a *kumhar* whose family had worked as *kammis* for Chowdri Abdullah's family. Given that these councillors depended on Chowdri Abdullah for either employment or patronage, they were unlikely to start questioning Chowdri Abdullah's decisions.

Although most of the other union councillors weren't kammis, tenants or permanent employees of Chowdri Abdullah's, all of them were his clients and power brokers. They were predominantly Jat smallholders. As power brokers they served as channels for people to approach Chowdri Abdullah for patronage. Kammis or smallholders who needed Chowdri Abdullah's patronage but didn't have close or direct relations with him would first approach one of them. Their knowledge of issues in their own neighbourhoods meant that they could keep Chowdri Abdullah informed about different problems around the union council. Their close knowledge of people's changing political alignments also meant that they were invaluable to his electoral strategies. Lastly, they also played an important role during elections because they could influence their neighbourhoods to vote for Chowdri Abdullah.

The issue of CCBs also clearly illustrates how devolution failed in its stated goal of empowering people and making politicians more accountable. The voluntary CCBs, described earlier, fell under the purview of the union council. With Chowdri Abdullah as union council nazim not only did no such voluntary organisation emerge but also almost no one that I spoke to was aware of what CCBs were supposed to be. Those few who were aware didn't see what the benefit of creating one could be.

In light of this scenario, it appears that the new local government structure simply extended elite factional competition into a new institutional set-up. Gondal landlords participated in it because they wanted to consolidate the power of their factions and undermine that of their rivals, not because they were concerned with accountability and participation. In what follows I explain why the devolution programme ended up reproducing landlord-dominated factional politics rather than enabling the emergence of more inclusive forms of political representation.

Who contests elections?

The August 2005 union council elections illustrate why elections in Pakistan, both under the devolution programme and more generally, failed to empower ordinary citizens. In Bek Sagrana the most obvious reason for this was that many kammis simply had to vote according to the dictates of the Chowdri on whose land they lived. However, even if this had not been the case, elections wouldn't necessarily have empowered people because the candidates who contested them belonged to the landed elite and therefore tended to represent its interests rather than those of ordinary people. In what follows I examine some of the factors that determined who contested elections.

As noted earlier, the union council election of August 2005 followed a five-year period where Chowdri Nawaz Ali's faction had been forced underground and where Chowdri Abdullah had become union council nazim for Gullhapur. In doing so, Chowdri Abdullah came generally to be seen as the most important and powerful politician in Bek Sagrana. By 2005, however, Chowdri Nawaz Ali and his faction had returned to the political scene and allied themselves with Chowdri Ghulam Ali Malkana. Given that Chowdri Haq Nawaz was the head of the Ghulam Baksh Ke faction in Bek Sagrana, and that he was the person in the union council of Daulatpur with the greatest interest in defeating Chowdri Abdullah, he was the most obvious candidate for the post of union council nazim in Chowdri Nawaz Ali's faction. The fact that Chowdri Haq Nawaz was a close relative of Chowdri Nawaz Ali also suggested him as an obvious candidate since this meant that he was more likely to share political and economic interests than someone more distantly related would. Moreover, his independent wealth meant that he would be able to raise sufficient funds to finance a significant proportion of electoral campaign costs, which could run up to and over Rs 800,000. Finally his ability to mobilise armed men also argued for his suitability to the challenge of taking on Chowdri Abdullah.

Nevertheless his potential candidacy also presented several disadvantages to the faction. Chowdri Nawaz Ali and his siblings realised that his bad reputation could lose them votes. Even though Chowdri Nawaz Ali and Chowdri Mazhar Ali knew that many people would still vote for Chowdri Haq Nawaz simply because he was their representative, they also realised the importance of enticing swing voters whose vote might be lost because they disapproved of Chowdri Haq Nawaz. Additionally, the fact that he had a criminal record rendered his application for candidacy unlikely to be accepted by the government, particularly since it was well known that he was aligned with politicians close to the opposition PML-N.[1]

Chowdri Nawaz Ali therefore had to find another candidate who would ideally be a wealthy close relative and who would additionally have his own connections with influential people. The candidate needed to be wealthy because this would enable him to fund his electoral campaign, and he needed to have his own connections because this would make him an effective patron. The first alternative to be considered was Chowdri Mahmood Abbas, who was Chowdri Nawaz Ali's nephew and the husband of Chowdri Haq Nawaz's sister. Not only was Chowdri Mahmood Abbas a close relative, but he was also a wealthy landlord and the son of the pious Sufi Ahmed Abbas, who was the second-largest landowner in Bek Sagrana after Chowdri Abdullah. Chowdri Mahmood Abbas was a lawyer who had studied at the Christian Foreman College in Lahore, the city where he continued to spend most of his time and where his two sons were being educated. He was well connected and had a reputation for being a *sharif admi*, or a pious and gentleman. Chowdri Mahmood Abbas, unlike Chowdri Haq

---

[1] Throughout Pakistan opposition candidates were barred from contesting elections because of criminal records as well as because of 'unspecified defects' in their educational certificates and of inadequate knowledge of Islamic injunctions.

Nawaz, had never touched a drop of alcohol, had never been involved in crime and remained relatively aloof from village strife; accordingly he had no criminal record and was not a particularly divisive figure. Furthermore, Chowdri Mahmood Abbas was likely to take up the offer of running for union council nazim because he bore a personal grudge against Chowdri Abdullah who had recently belittled him in public.

The fact that Chowdri Mahmood was a sharif admi had its disadvantages, however. As noted, the advantages included the fact that he wasn't a divisive figure and that few people, other than possibly Chowdri Abdullah, had any significant quarrel with him. His status would also provide a cover of respectability for Chowdri Haq Nawaz who would continue playing a prominent role in the Ghulam Baksh Ke faction. The disadvantage was that being a sharif admi could suggest that a person was neither comfortable nor familiar with the political utility of violence and force, and would therefore be ineffective in a political context that required leaders who were ready to face up to powerful opponents by these and other means. What's more? Chowdri Mahmood was widely perceived to be lazy. Both villagers and relatives pointed to his disregard for his farm, and the fact that his labourers did as they pleased and frequently stole from him, as evidence for his laziness. Even his father once publicly reprimanded him after Chowdri Mahmood left his wheat standing several weeks after everyone else had harvested theirs. His father called him a *badshah admi*, a term indicating a person who lived a kingly lifestyle and who had little concern for work. As a result of the widespread perception that he was lazy, several people in the Ghulam Baksh Ke faction quietly voiced their concern about Chowdri Mahmood's effectiveness as a candidate.

The most reasonable alternative to Chowdri Mahmood who was willing to contest the elections against Chowdri Abdullah was Dr Shafique Gondal, a distant Gondal cousin of Chowdri Haq Nawaz who owned land in Bek Sagrana. Unlike Chowdri Mahmood who came from a higher-ranking family of Gondals, Dr Shafique's family came from the Gondals' middle ranks. Nevertheless, Dr Shafique's father owned some 15 acres of citrus orchard that had permitted him to send all of his three sons to university in Sargodha. The eldest of the three was now a police inspector working in the office of the Assistant Superintendent of Police (ASP) in Sargodha, and the youngest worked at the government education board in Sargodha. Dr Shafique, the middle son, worked in the Sargodha government health department. The three siblings lived in separate rented accommodation in Sargodha and occasionally travelled to the village on their motorcycles. All three brothers had a good reputation and there was nothing in Dr Shafique's personal track record that would have allowed the government to interfere with his nomination. Additionally, although Dr Shafique and his brothers were not as wealthy or as well connected as Chowdri Mahmood or Chowdri Haq Nawaz, they were still in a position to pull strings for their relatives and to dispense patronage to villagers. Dr Shafique's younger brother (who worked in the Sargodha Education Board) had, for example, obtained a job for the son of the Imam Masjid as a school teacher in his home village. He had also helped another villager obtain a high school matriculation certificate even though the man in question had never completed high school. Dr Shafique's elder brother (who worked at the office of the ASP in Sargodha) had helped a few villagers in sorting out problems with the police. Dr Shafique himself had helped obtain hospital treatment for several people in Bek Sagrana and the surrounding area and was promising to use his position to secure government jobs as low-ranking hospital *chowkidars* and cleaners for others.

Despite his relatively solid position in the village Dr Shafique had neither the financial means nor the power to stand independently of either Chowdri Nawaz Ali's or Chowdri Abdullah's faction. As discussed earlier, election costs could run up to Rs 800,000 and beyond, covering everything from fuel for vehicles to payment for votes, and the funding of minor infrastructural projects to payment for meals and campaign advertisements. Funds were also required to provide poorer voters with transport to the polling station on election day. In addition, Dr Shafique was likely to be thought of as a weak candidate by the electorate if he didn't have the backing of influential politicians. As noted, the electorate expected and wanted powerful and well-connected politicians since they were the people who could deliver on promises of patronage. Furthermore, especially against a figure as powerful and well connected as Chowdri Abdullah, a candidate who wasn't well connected risked being intimidated and victimised through the threat of force and the use of the judiciary and its processes.

Thus, Dr Shafique's candidacy was realistic only because, after much deliberation, Chowdri Nawaz Ali and his close allies decided to support him. The campaign expenses were principally going to be covered by a collective effort of the principal stakeholders including Chowdri Nawaz Ali and Chowdri Haq Nawaz, as well as by occasional contributions in the form of a vehicle or a meal by other supporters including Chowdri Mahmood . The latter, for example, put his car at the disposal of Dr Shafique for more than 10 days and paid for a great deal of the fuel expenses incurred by the extensive daily travel to various households and villages in the union council of Daulatpur by the candidates and their close supporters. Nevertheless, the principal support for Dr Shafique came from Chowdri Haq Nawaz who was constantly at his side throughout his campaign and who put his vehicle, funds and gunmen at Dr Shafique's disposal. It is worth noting that despite being supported by powerful politicians, many villagers remained sceptical of Dr Shafique's capacity to deliver patronage. They claimed that Dr Shafique was a nobody, and they underlined their view by pointing to the fact that he rode around on a motorcycle while his opponent, Chowdri Abdullah, rode around in a brand-new Lexus Land Cruiser.

The foregoing makes it clear why ordinary villagers didn't present themselves as candidates in the union council elections. Given what it took to qualify as a candidate in both practical and reputational terms – even to the extent that Dr Shafique, despite his powerful backing and his already-respectable personal status and contacts, remained a figure of questionable political effectiveness to ordinary villagers – the thought of a candidate emerging from the ranks of kammis or labourers was unthinkable and literally laughable. Muhammad Hussain, discussed in the previous chapter, explained to me that people's mentality was such that if a kammi stood up to contest elections everyone would laugh at him. Not only would the chowdris laugh at him (and possibly punish him for insubordination), but his fellow kammis would also find him ridiculous. Hussain said that people would completely discount a poor politician who didn't campaign with a large convoy of cars and cohort of gunmen.

Hussain concluded that the reason that the poor were oppressed and powerless was the result of their mentality, which worshipped the outward signs of power and wealth. Yet the poor had good reason to believe that only the rich and powerful could contest elections and become effective politicians. Because politics was largely about power and patronage, it was obvious to people that a poor kammi, even in the unlikely event that he managed to be elected as union council nazim, couldn't deliver patronage to supporters in the same way that a Gondal Chowdri could. One reason for this being that most kammis were illiterate and would therefore be unable to deal with the bureaucratic paperwork that characterised most encounters with the state. Another reason was that unlike the Gondals, kammis didn't have the social capital in terms of connections with influential officials that would allow them to deliver patronage.

Campaigning

For both Chowdri Abdullah and Dr Shafique the key to winning the union council elections was their capacity to gain the allegiance of the various heads of zamindar families living in the 13 villages that comprised the union council of Daulatpur. By gaining the allegiance of these zamindars, the majority of whom were middle-ranking and owned less than 10 acres of land, the candidates sought to win not only the votes of entire households (which included close servants), but also the votes of these zamindars' dependent and client households. It was generally the case that members of zamindar households voted according to the decision of their household's head. Where heads of households had conflicting loyalties or desired to remain neutral they either told their household members to abstain from voting altogether or instructed them to split their votes equally between the candidates. Zamindar household heads could also determine the way their tenants, labourers and kammis voted. As a result, campaigners tended to bypass kammis and labourers who were known to be under the authority of a particular chowdri and simply approached the latter for their votes instead. In fact directly approaching kammis without first consulting the chowdri under whose authority they lived was likely to be interpreted as a challenge to his authority.

In order to obtain the votes of these zamindars, candidates relied upon both friendship and patronage. Candidates could rely upon their close supporters to mobilise their zamindar friends on their behalf. In cases where the core faction members had neither kinship nor friendship ties with particular zamindars, these zamindars had to be won over either with promises of future patronage or, more effectively, by offering them immediate patronage. Thus, approximately two months before the elections candidates from both sides could be found busily travelling around the district trying to resolve different zamindars' issues in order to gain their votes. Among other things, candidates became involved in resolving land disputes, mediating with the police, getting people treated in hospital, getting school teachers to attend to their duties, and repairing municipal infrastructure such as roads, bridges, phone lines, and gutters.

Although Dr Shafique was able to resolve several police disputes thanks to his elder brother's help and was also able to get some people who needed treatment admitted in hospital, Chowdri Abdullah had a significant advantage when it came to granting patronage for two obvious reasons: because he was the incumbent union council nazim and because he was aligned with the ruling coalition. In one instance, for example, some Ranjha Jat cultivators who lived along the irrigation canal downstream from Daulatpur offered to give their votes to whichever candidate built a bridge over the irrigation canal that separated them from the main road. Although several bridges already existed, the cultivators complained that they had to walk thirty minutes upstream to get to the closest one and that this caused them unnecessary delays. Dr Shafique didn't have the Rs100,000 needed to build the bridge at his immediate disposal, but he promised the cultivators that he would get back to them with the funds within a week. However, before a week had passed, Chowdri Abdullah, who had access to local government funds, appeared in the hamlet and offered the cultivators a cheque for Rs100,000 on the spot, but on condition that they all went into the local mosque and made a vow to give him their votes. The cultivators happily accepted the offer, and Dr Shafique lost out because he had been unable to muster funds quickly enough.

Similarly Chowdri Abdullah managed to snatch a number of zamindar votes away from Dr Shafique by getting the authorities to install a telephone line for them. The zamindars in question, who lived downstream from Bek Sagrana, had tried to get the relevant authority to install a phone line for years but had failed to make any progress. With the onset of elections Dr Shafique had promised to help them. However, Chowdri Abdullah managed to get the relevant authorities moving before Dr Shafique was able to do anything, and he accordingly gained their votes. Apparently Chowdri Abdullah achieved this by getting MNA Khuda Baksh Mekan to pull strings in the relevant ministry, providing another example of the power of connections in political outcomes.

Although the bulk of patronage was dispensed to existing and potential supporters from zamindar families, in certain instances both factions did try to reach out directly to kammis and landless labourers. This was generally done in cases where these people were either not under the direct authority of any particular chowdri or where the chowdri under whose authority they were desired to remain neutral was indifferent about the elections. People who weren't directly under the influence of particular chowdris included some who lived in the Bhutto colonies, some who lived in villages and who earned an independent livelihood from the chowdris and some who had settled on government land adjacent to paved roads. Unlike people who settled around the farmhouses (*deras*) of chowdris or in a house on their fields and who were therefore generally supposed to vote for whomever their chowdri told them to vote for, these people were largely free to cast their vote however they pleased.

In most cases these independent voters were cynical and indifferent about the elections since they believed that they had little to gain from them. Many kammis that I spoke to during the campaign period told me that chowdris who had a stake in the elections suddenly became friendly and greeted them as they drove past. One day a Mussalli who lived in a Bhutto colony and earned his livelihood independently of any Gondal chowdri, told me that one of the Gondals campaigning with Dr Shafique even got out of his car to greet him and ask him for his vote. He commented that this was unthinkable outside of electoral campaigning when if a Gondal addressed him it would be either to boss him around or to insult him. Like many others he told me that as soon as the campaigning was over things would revert to the old way.

Additionally poor independent voters were often indifferent because they feared that taking sides in the elections would bring them problems.

Even if they lived in colonies or on government land next to the road they couldn't afford to antagonise landlords in their neighbourhood who might prevent them from using their orchards to collect firewood and to go to the toilet or deprive them of future patronage or, in the worst of cases, evict them from their homes. Many therefore kept quiet about their electoral preferences, even if they had strong ones. Keeping quiet kept them safe from informants wanting to ingratiate themselves with the landlords by informing them about peoples' voting intentions. Many preferred to avoid all trouble by simply not voting. They not only feared that their voting intentions might be revealed but that they might also be found out on polling day when the secrecy of the ballot box was far from adequately ensured.

The most effective way to secure swing voters was with cash. Dr Shafique's limited funds meant that he was largely unable to do this, but Chowdri Abdullah secured a number of votes two days prior to the elections by paying people Rs 500. Several people including Dr Shafique commented that by buying people's votes Chowdri Abdullah was effectively carrying out a commercial transaction that absolved him of any future responsibility towards his voters. Dr Shafique explained that in future if anyone who had received money for their vote went to Chowdri Abdullah for patronage they would be told that they had already received money and be turned away. People realised this, but when queried about it many said that cash in hand was better than promises which were likely to remain unfulfilled. Dr Shafique could only offer promises while Chowdri Abdullah had something to offer immediately.

Rigging

The rural poor also regarded elections cynically and indifferently because they were almost invariably rigged. Their cynicism was vindicated during the 2005 union council elections. These elections, like the previous ones in 2001, were marred by administrative mismanagement and rigging. The ICG (2005) reports that polling staff were inadequately trained, that the secrecy of polling booths wasn't properly ensured, that there was a shortage of ballots and indelible ink, and that outdated voting lists were used. They also report the widespread incidence of ballot stuffing, multiple voting and vote buying. Polling booths were placed in pro-government villages or pro-government wards of villages in order to intimidate opposition voters and politicians out of voting. In many instances votes were discounted and names were crossed off voting lists. Voting by women was particularly marred by irregularities because they were allowed to use old identity cards with thumbprints rather than photographs thereby allowing identity fraud during the polling. In the North West Frontier Province and in certain

Punjab districts women weren't allowed to vote by leaders and by their male kin. In two polling stations in Sargodha district family members of PML-Q candidates were found stamping ballot papers for allegedly old and illiterate women.

It goes without saying that most of the rigging took place on behalf of PML-Q candidates, but mismanagement of the polling process, understaffing and corruption within the Electoral Commission of Pakistan appear to have allowed at least a few opposition politicians and their supporters to rig the polling in their own favour.[2] Generally however election staff and the police overlooked and assisted the rigging process in favour of the ruling party. The police harassed opposition candidates, allowed the seizure of polling stations by pro-government groups and allowed people to bear arms around the polling stations. The police and security services were particularly harsh towards PPP candidates and supporters in Sindh. Throughout Pakistan serious irregularities were also reported in the process of counting the votes. In the Punjab the worst rigging took place in the Punjab chief minister's home district of Gujrat – to the north of Sargodha.

A number of these pre-poll, poll day and post-poll irregularities took place in Bek Sagrana. Chowdri Abdullah had a distinct advantage when it came to both pre-poll rigging and polling day rigging because he was the incumbent union council nazim and that he was aligned with the ruling PML-Q coalition. However, this didn't render his opponents into mere victims of the government: they too partook in electoral malpractice. The printed press and international observers, including the ICG, have rightly tended to emphasise the rigging carried out by Musharraf's government. However, this didn't make opposition politicians paragons of democratic virtue. Another problem with focussing almost solely on rigging by pro-government supporters is that doing so seems to assume that Musharraf's military-backed coalition had the monopoly of power whereas in fact opposition politicians could often wield their own power against the ruling coalition by virtue of their local strangleholds and influence at various levels of government.

Chowdri Abdullah's access to government funds was one among many of the means at his disposal to carry out pre-poll rigging. Additionally, as the incumbent pro-government union council nazim, Chowdri Abdullah had an important degree of influence over the issuing of identity cards, which were necessary for voting, and enjoyed a significant measure of support from

---

[2] In addition to being biased an ICG report of 2011 describes the Electoral Commission of Pakistan as 'poorly managed, inadequately resourced, under-staffed and under-trained' (ICG 2011: i).

the local police. Chowdri Abdullah obtained identity cards for supporters who didn't have one yet, and created hurdles to obtaining them for people whom he knew or suspected would vote against him. Dr Shafique alleged that Chowdri Abdullah's faction managed to obtain up to 200 fake identity cards for supporters. Although this figure may not have been accurate, there is no doubt that fraudulent identity cards were issued. This was accomplished by having identity cards issued in the names of people who were on the electoral register but who had died and had not yet had their names removed from it. Chowdri Abdullah gave these cards to supporters who, for whatever reason, were not on the electoral register. However, in some cases they were given to supporters who already had one so that they could vote twice.

Chowdri Abdullah also managed to use police support to undermine his opponents' chances of electoral success. One of the ways in which he did this was by using them to entangle his opponents in fabricated cases of abduction and theft. This was not only an effort to make his opponents lose money and time extracting themselves from the police investigations (and possibly even to eliminate them completely from the electoral contest), but also to discredit them in the eyes of the electorate for their apparent involvement in criminal activities. On one occasion Chowdri Abdullah managed to get opponents detained for the illegal possession of weapons as they were driving towards a political rally to be held in the village of Bek Sagrana. Chowdri Abdullah had heard about the rally and had called upon his contacts in the local police to stop them and check their cars for unregistered weapons. Far from unusual, the possession of unregistered weapons by politicians was commonplace, and given that his opponents, who were duly stopped at several police checkpoints, did have unregistered weapons with them, they were detained by the police for several hours and thus prevented from attending the rally. Chowdri Haq Nawaz and Dr Shafique were quick to conclude that Chowdri Abdullah had been behind the actions of the police since having three of their cars stopped separately on the same day was unlikely to be merely coincidental. The fact that I later heard a close supporter of Chowdri Abdullah boasting about how they had got their opponents into trouble with the police lends credibility to Chowdri Haq Nawaz and Dr Shafique's conclusions regarding Chowdri Abdullah's involvement in the affair.

A few days after the unregistered weapons episode, Chowdri Abdullah allegedly tried to get Chowdri Haq Nawaz entangled in a case of buffalo theft. One night, one of Chowdri Haq Nawaz's gunmen, who was actually covertly working for Chowdri Abdullah, stole two buffaloes and placed them in Chowdri Haq Nawaz's dera. Having done this the gunman fled to Chowdri Abdullah's dera under cover of darkness. The next morning before dawn the police, acting on a tipoff, raided Chowdri Haq Nawaz's dera to retrieve the stolen buffalos and arrested two of his servants. The two servants appeared not to know anything about the stolen buffalos and claimed that the first time they had seen them was that morning. Chowdri Haq Nawaz happened to be away that night but when he heard of the situation he appeared genuinely surprised but it took him little time to conclude that Chowdri Abdullah had framed him. This was later confirmed when he was summoned to the police station where the Station House Officer (SHO) wanted to file an FIR against him. Although the SHO was rumoured to be in cahoots with Chowdri Abdullah, another officer who was on good terms with Chowdri Haq Nawaz told him, in my presence, that the person who had called to inform them about the stolen buffalos in his dera was none other than Chowdri Abdullah. Police officers often kept good relations with local leaders regardless of whether they were in the opposition because they knew that the opposition would one day regain power and that they might be transferred if they had been high-handed or uncooperative. In the end although the SHO threatened to place an FIR against Chowdri Haq Nawaz, this never happened because Chowdri Nawaz Ali was able to use his own contacts higher up in the police hierarchy to prevent it.

Nevertheless, Chowdri Abdullah somewhat succeeded in discrediting Chowdri Haq Nawaz in the eyes of the public because the case was published in the Sargodha press; but only somewhat because most villagers didn't trust or even read the papers. In any case it didn't really matter because everyone knew that both Chowdri Haq Nawaz and Chowdri Abdullah were involved in trafficking stolen livestock and goods.

The foregoing examples illustrate well the chess game played by politicians at election time, marshalling their respective resources in the form of influence and contacts both offensively and defensively in a variety of shady gambits. In the buffalo episode, Chowdri Nawaz Ali was able to mobilise some of his own contacts to counter Chowdri Abdullah's move against him. Later he was also able to pull strings with the local judiciary in order to obtain favourable results during the polling. The fact that Chowdri Nawaz Ali was a lawyer and had various friends in the local judiciary meant that he was able to influence where the polling stations would be located. Although in the union council of Daulatpur he was unable to get the polling stations to be located in friendly territory he was at least able to ensure that they weren't located in enemy territory. Thus polling stations were located where the supporters of both factions were roughly equal in numbers and strength, and that no one faction would be able to takeover the polling station wholesale.

In at least one instance Chowdri Nawaz Ali's contacts in the judiciary got him a polling station in his home village as well as a favourable polling officer prepared to overlook an array of electoral malpractices in his faction's favour. The village of Bukhuwala – Chowdri Nawaz Ali's village of residence – wasn't part of the union council of Daulatpur but was in one of the many union councils that Chowdri Ghulam Ali Malkana and Chowdri Nawaz Ali sought to capture. In Bukhuwala the electoral contest was between Chowdri Nawaz Ali's candidate – who belonged to the larger faction in the village – and a Makhdoom from a neighbouring village who was part of the wider pro-government political block that included Chowdri Abdullah and the MNA Khuda Baksh Mekan. The Makhdooms should have had the advantage since they were with the ruling party, but the Gondals discretely rigged things in their own favour. The polling officer deceived the Makhdooms into thinking that he was working for them when he was in fact overlooking multiple voting in favour of Chowdri Nawaz Ali's candidate.

Polling day

From early morning on the day of the elections both factions were busy coordinating buses to transport supporters to the polling station. Chowdri Abdullah's brother used the bus he owned to bring supporters back and forth between their villages and the polling station. Chowdri Abdullah also secured another bus through the government. When it came to fetching more influential voters from zamindar biraderis, Chowdri Abdullah dispatched his Lexus Land Cruiser and his brother's new Toyota Corolla.

Dr Shafique had no buses at his disposal, but he gathered a number of vehicles from friends and supporters, including two pickup trucks. Around the polling station, which was situated inside the village school, both factions set up large tents where people were offered chicken, roti and sweet rice as well as cold lemonade (*skanjbi*) made with local citrus as an added incentive to come and vote. By looking at who turned up to their tent, candidates were able to monitor people's loyalties (see Wilkinson and Kitschelt on how politicians monitor voters).

The tents were reminiscent of wedding tents and gave the polling a festive feel. For many locals who had little to gain or lose in these elections, polling day was seen as a festive occasion where they could get free food and drink and observe and gossip about village leaders. Gossip ran from speculations about their chances of success in the polls to the various conspiracies surrounding the elections, as well as to the various quarrels between members of the two opposing factions in the preceding weeks. Many people, particularly the young, also looked forward to a bit of entertainment in the form of a fight during the polling. For many of the poorer villagers and children who weren't even going to vote this was an occasion to eat as much as they could, and some of them even ate and drank from the two enemy tents.

After weeks of campaigning and entanglements with the police as a result of Chowdri Abdullah's Machiavellian schemes, Chowdri Haq Nawaz' nerves were frayed, and on the night prior to polling day he repeatedly declared that if Chowdri Abdullah continued rigging the elections and harassing him and his men he was going to take him on personally. He even declared that only one of them would be left alive by the end of the elections because it wasn't possible for the two of them to coexist in this world. In the end neither of them lost their lives, but there were many moments throughout the day when irregularities at the polling stations almost erupted into gunfights.

The inside of the polling station was very disorderly, and there were many people loitering in it who weren't supposed to be there. Many were there to unofficially assist the polling agents – representatives of both candidates – there to put a check on rigging by rivals and to ensure that the polling staff and the police were acting impartially. They were also there to – illegally – fill in the voting papers of elderly supporters who needed assistance because of weak eyesight, or because they didn't understand how to fill in their ballots, as well as to fill in the ballots of people who had pledged their votes. In the latter cases polling agents and their assistants were thereby able to determine whether people who had pledged their votes actually delivered them and therefore whether they were entitled to future patronage. Whenever the rival polling agent and his assistant weren't vigilant, they also intimidated and tried to fill in the ballot papers of people who hadn't pledged their vote to anyone. I heard of a couple of occasions when they grabbed people's ballot papers and filled them in themselves.

When rivals quarrelled it was generally because one party's supporter had seen a rival supporter doing this type of thing. At one point one of Dr Shafique's supporters saw one of Chowdri Abdullah's supporters whispering something into a kammi's ear – presumably telling him how to vote – and then looking over his shoulder to make sure that he had done as told. Seeing this Dr Shafique's supporter complained and rival supporters started hurling accusations at each other and threatening to draw their guns. The quarrel ended only after the village imam interfered and made all the parties involved hold each other's hands and raise them in unison while reciting the Islamic profession of faith (*kalma*). Chowdri Haq Nawaz, however, was left muttering about how by the end of the day either he or Chowdri Abdullah would be no more.

In addition to the close supporters of each party there were plenty of other random people who tried and often succeeded in casting their votes several times. A young man told me that he had voted four times and another told me that he had cast his vote twice for Chowdri Ehsanullah and twice for Dr Shafique just for the heck of it. A man who wasn't even on the electoral register boasted about how he had voted three times in favour of Dr Shafique. Informants later told me that one of the things that facilitated multiple voting was that ballots weren't numbered so that once people were inside the polling station they could pick up and cast as many ballots as they wished, and also that there was no way of telling how many times an individual may have voted after the ballot boxes were opened.

Understaffed, polling and police officers couldn't have prevented these irregularities even if they had been determined to do so. It was impossible for them to control either the flow of people in and out of the polling station or what all these people were doing inside it. Moreover many polling officers were afraid to raise objections about irregular voting because they didn't want trouble with the armed supporters of either candidate. Once in a while police officers, who spent most of their time in Chowdri Abdullah's tent eating and drinking, would barge in and unceremoniously expel some of the excess people inside the polling station. After a while, however, chaos would return to the polling station and the police would eventually repeat the same exercise all over again.

After the voting had been completed by mid-afternoon, the hard-core supporters of each faction stayed on around the polling station until nine o'clock in the evening when the results were released. Although the results were never officially published in the village, Chowdri Abdullah appeared to have won by a comfortable margin in the Union Council as a whole. In Bek Sagrana the contest must have been relatively close despite police partiality towards Chowdri Abdullah due to the roughly equal representation of both factions during the polling. The moment the results came out the supporters of Chowdri Abdullah began shooting rounds into the air with their Kalashnikovs and taunting the supporters of Dr Shafique. That night celebratory shooting was heard until late, and over the next few days samosas and jalebis were distributed to supporters who went to congratulate (*mubarak dena*) Chowdri Ehsanullah and Chowdri Abdullah on their victory.

There were no celebrations on Dr Shafique's side although they comforted themselves with the idea that their faction was the more popular one and that Chowdri Abdullah had only won thanks to fraud. They alleged that he had played all sorts of tricks on them, purchased votes, tampered with the voting lists, and issued hundreds of fake identity cards. Chowdri Haq Nawaz was unable to contain his bitterness, and on the day after the elections he threatened to ban the Rajputs from using their community centre adjacent to his land because they had voted for Chowdri Abdullah. The fact that the secrecy of the ballot was not ensured meant that it did not take a long time for the chowdris to ascertain who had voted for whom. Chowdri Haq Nawaz was particularly angry with the Rajputs because he claimed that Chowdri Nawaz Ali had given them their community centre and had even provided some of them with government jobs. Following Chowdri Haq Nawaz's threat of eviction one of the Rajputs appealed to one of Chowdri Haq Nawaz's close supporters known to be a level-headed man in order to stop Chowdri Haq Nawaz from taking action against them. The chowdri had reasoned with Chowdri Haq Nawaz and had eventually convinced him that he should be merciful, in part because evicting the Rajputs would only lessen his chances of winning future elections.

Although Chowdri Nawaz Ali's candidate was able to defeat the Makhdoom candidate in his home village of Bukhuwala (partly thanks to the friendly polling officer), it also became clear on the day following the elections that the alliance which included Chowdri Abdullah had managed to win the majority of union council seats in the area. This meant that they would be able to gain a district nazim favourable to their interests. Even though Chowdri Nawaz Ali vowed to contest the results through the local courts on the basis that they were the result of rigging, he and his allies would have to wait until 2008 in order to regain some of the political power that they had lost since Musharraf's takeover.

Conclusions

As in Bek Sagrana, PML-Q candidates swept the polls in most villages in the Punjab. Even more importantly for the Musharraf government, it was subsequently able to obtain all but a handful of district nazims seats in the Punjab. The ICG reported that in 2005 on election days alone at least 60 people were killed and some 550 people were injured across Pakistan (ICG 2005: 10). In light of this, and of the evidence presented in this chapter, to call the results a victory for enlightened moderation as the government did would be disingenuous. The PML-Q's victory in the polls was merely the victory of those with more power over those with less.

While poll rigging was certainly a problem, it isn't clear that free and fair polling would have produced genuinely representative local government bodies. As shown earlier, bureaucratic connections coupled with wealth and an ability to mobilise the use of force were the key determinants of electoral success. This meant that the common man, for whom the local government system was allegedly designed, was unlikely to ever get voted into office. Moreover the presence of powerful political brokers – like Chowdri Abdullah and his rivals – limited people's opportunities for meaningful political participation; their dominance meant that the poor had fewer opportunities to complain about government services or become involved in politics even at the local level. Moreover even without poll rigging the elections were never going to be entirely fair because pro-government politicians would always get more government resources than opposition politicians and therefore would have better chances of getting voted. Villagers made this clear by telling me that voting for Dr Haq Nawaz was pointless because he was in the opposition and wouldn't be able to do their 'work' (*kaam*) for them. Thus even without poll day rigging, Musharraf's government is likely to have won a majority of local government seats – although it is unlikely to have gained almost near monopoly over district nazim seats as it did.

One lesson to be drawn then is that the devolution programme focussed too much on the institutional impediments to democracy and overlooked the social ones. Most notably, it failed to take into account how landed power might undermine attempts at grassroots governance. Similarly to focus solely on fixing procedural and technical aspects of elections – such as preventing multiple voting and booth captures – as the international community did during the various elections held under the Musharraf regime would likewise be to overlook the broader impediments to democratisation in Pakistan (see ICG 2011). It would be to overlook a larger picture in which government is systematically biased, both during and beyond elections due to military interference in politics. Throughout Pakistan's history military regimes have used institutions such as the judiciary, the police and the Electoral Commission of Pakistan to benefit its allies and supporters. Even between 1988 and 1996 when Pakistan was formally democratic, the military used bribery, coercion and electoral manipulation to control political outcomes. Such meddling resulted in political instability, the erosion of law and order and stunted Pakistan's democratic development. Moreover, as argued in chapter four, such meddling further entrenched the landed elites within Pakistani electoral politics.

Following Benazir Bhutto's murder the military appears to have reduced how much it meddled in politics, and there are signs that Pakistan democracy may finally be given a chance to start maturing. To start with, Musharraf feared popular unrest following Benazir Bhutto's murder – in which he was widely seen as involved – so he only selectively rigged the 2008 elections. Interestingly an informant who was a polling officer in a village near Bek Sagrana told me that supporters of the local PPP candidate and not those of the PML-Q candidate captured the local polling station and stuffed the ballot box. Apparently the polling officers didn't have enough police protection, and feared harassment if they raised objections to the rigging. My informant said that only the military could have prevented rigging by heavily armed party workers. What is certain is that to prevent this, as well as to prevent the type of polling station chaos described in this chapter, electoral procedures need to be significantly improved. For this to happen, the military will have to permanently stop meddling and thus weakening institutions such as the ECP, the police and the judiciary.

6
ISLAM, SELFLESSNESS AND PROSPERITY

There aren't any. Not real peasants. Not revolutionary peasants. The ones I have met are fighting for their feudal lords, not against them. They are fighting to preserve the status quo. They are fighting so that their feudal lords can keep them in their shackles. They are subverting the genuine class struggle of workers like me and you.

(Hanif 2008)

The equation of Islam and the central government encouraged arguments, some more pious and others, about Islam's potential leavening effects on all forms of social divisiveness. Its categorical pronouncements on the right to own private property, matched by a very broadly defined concept of social justice, could pre-empt moves by ungodly communist ideologues to promote class conflict. For the landlords of west Pakistan, anxious to cut short all moves towards agrarian reforms, for the trading and commercial groups, determined to make a killing without let or hindrance, for the urban propertied classes, looking to secure and augment their wealth, and for a state needing to consolidate and expedite the processes of capital accumulation, Islam offered a moral escape from one too many awkward realities. As Liaquat Ali Khan pontificated, the people of Pakistan should 'follow the teachings of the Prophet and not those of Marx, Stalin or Churchill[sic].

(Jalal 1999)

A very brief history of Islamisation in Pakistan

While previous chapters have focussed on the political and material aspects of landlord dominance, this chapter focuses on its ideological aspects, looking at how Pakistan's espousal of Islam as the national ideology served the interests of its dominant classes. Since Pakistan's birth in the wake of the traumas of partition in 1947, one of the dominant concerns of its ruling classes has been to create national unity through Islam. Islam was deployed to keep together a culturally diverse population divided by extreme economic disparities. This quest for unity through Islam purposely diverted political attention away from the need for greater social justice and democracy. During the cold war it even served to actively combat leftist groups seeking greater social justice. Under General Zia ul Haq, Islam was specifically used to combat the left and invoked to reverse land reform legislation initiated by Bhutto. During this same period left-leaning student groups virtually disappeared from university campuses and were replaced by the Islami Jamiat-i Tulabah, the student wing of the Jamaat-i-Islami. Also, because at independence it had lacked a solid constituency, the Muslim League deployed Islam to give itself and the central government which it dominated legitimacy.

Jinnah himself, despite his secularist outlook, was quick to 'take refuge in Islam to survive the cross-fire of provincialism and religious extremism' (Jalal 1999: 280). Unwilling to concede greater powers to centripetal provincial forces, Jinnah's central government sought to establish itself as the guardian of the faith for which so many had sacrificed their lives at partition. Forces pressing for greater provincial autonomy and cultural recognition, particularly in East Pakistan and Baluchistan, were quickly accused of seeking to undermine the sacrifices made for a Muslim homeland. From Pakistan's earliest days the central government accused its critics of being Indian agents working against Islam – a claim that has ever since been used by the establishment as well as by anyone wishing to discredit political opposition and dissent.

Lacking significant measures to address economic and power imbalances between regions or social classes, government-led Islamisation neither gave the central government greater legitimacy nor fostered a sense of national unity. This became abundantly clear with the secession of East Pakistan in 1971 and with growing – and continuing – unrest in Baluchistan over provincial autonomy and shares in gas revenues. Without an institutional, democratic framework to facilitate the peaceful settlement of grievances, and with the often violent suppression of dissent, individuals increasingly took recourse to violence and agitation in order to redress their grievances.

By seeking legitimacy through Islam, the central government cleared the way for a wide array of Islamic ideologues to make their bid for power as legitimate opposition. Since Islam in Pakistan was never the monolithic identity that the ruling classes sought to portray, it was unlikely to succeed in creating social unity on its own. Not only did Pakistan's citizens belong to a variety of different Islamic traditions, including various different Shia and Sunni schools of thought, but those traditions themselves varied not only according to region but also, as this chapter will show, according to class. Thus Islam soon became the main point of contention between Islamic scholars from various traditions, all of whom vied to determine the extent to which the laws of the state were in accord with the teachings of Islam. Moreover, as the work of anthropologists such as Marsden (2005) illustrates, Islam came to dominate everyday public debate among ordinary people too.

Because the education system was plagued by corruption and teacher absenteeism, the rural-and urban-poor increasingly flocked to often better-funded religious schools. In the Punjab, the role of madrassas was particularly prominent in the South where feudal landlords opposed popular education fearing that it might deprive them of labour and make people more aware of their rights and more assertive. According to one study conducted in 1998 there were 5,000 'ghost schools' in the Multan division and 800 in the Sargodha division (Abou Zahab 2002: 82) because landlords used them as cowsheds to prevent people going to school.

On the other hand functioning Madrassah's were plentiful: there were 2,512 of them in the Punjab and 1,619 in the Southern Punjab alone. They were attractive to the poor because they were cheap but also because they offered the possibility of status and even power. A religious scholar or a Jihadi – who might become a martyr (*shaheed*) – in the family could significantly raise its social status. Moreover a religious scholar or Jihadi belonging to an armed militant outfit could even induce fear and respect in landlords. The families of martyrs were also likely to receive money from militant outfits for their son's sacrifice (*qurbani*).

While madrassas sometimes played an important social welfare function, they could also sow sectarian hatred and conflict. Religious preachers at some madrassas spent a great deal of time denouncing the heresies of rival schools of thought – not to mention those of Christians and Hindus. Some Deobandi Sunni preachers were vociferous in denouncing members of the Ahmedi sect as non-Muslim because of their alleged denial of the finality of Prophecy with the Prophet Muhammad (SAW), and they succeeded in having them officially declared as such by the Pakistani state in 1974. The same religious leaders, particularly activists from Sipah-i Sahaba, turned upon the Shias and sought to extend the anti-Ahmadi legislation to them, particularly during the Islamisation campaigns of the Zia era when Shias started pressing for separate religious education and objected to the imposition of a uniform Islamic tax (zakat). The result has been extensive violence between Shias and Deobandis with estimates suggesting that more than 2,000 people were killed, with thousands injured and maimed, during incidents of sectarian violence from 1985 to 2005 (ICG 2005b:1).

The central Punjabi district of Jhang, just south of the Sargodha district, was particularly hard hit by sectarian conflict. There the followers of preachers from rival Deobandi and Shia sects clashed in turf wars for the control of neighbourhoods and mosques. Deobandis also occasionally clashed with Barelvis whose style of saint-worship they denounced. Sectarian conflict in Jhang also appears to have channelled both rural and urban grievances against the predominantly Shia rural landlords of the area. Sectarian leaders like Maulana Haq Nawaz Jhangvi believed that the powerful Shia landlords that dominated the politics of Jhang were not only exploiting the rural population with their economic and political power but also with their spiritual influence, because many of them were also pirs. Jhangvi believed that these Shia magnates were keeping the rural population ignorant of true Sunni Islam and even gradually converting it to Shiism. Sipah-e-Sahaba (which Jhangvi had founded) wished to further the interests of the predominantly immigrant Sunni merchant population against the political power of the wealthy Shias. The principal aims of the heavily armed Sipah-e Sahaba were to combat the local influence of the Shia Magnates and also, more broadly, to ban Muharram processions, have the Shias declared a non-Muslim minority and make the Sipah-e Sahaba brand of Sunni Islam the official religion of the State.

Particularly relevant to this book is the fact that among the supporters of Sipah-e-Sahaba were people who had migrated to the cities from the countryside and who had experienced the high-handedness of landlords (see Zaman 2002:125). According to Zaman these people were particularly attracted to calls by Sunni preachers urging them to fight feudal oppression and 'the ideological legitimation offered for it' (ibid.). Segments of the urban population were also attracted to Sipah-e-Sahaba because of its appeal to the interests of the common man in 'an urban milieu where administrative and judicial authorities are inefficient and corrupt, and are widely held to act in concert with the landed elite' (ibid.). Jhangvi himself, a man of humble rural origins, is alleged to have spent time looking after the welfare of the poor. He was particularly noted for having spent time at government courts and police stations helping poor illiterate litigants whose chances of a favourable outcome were limited without the help of a patron. Similarly the Tehrik-i-Nifaz-i-Shariat-i-Muhammadi (TNSM), otherwise referred to as the Swat Taliban, channelled popular grievances by decrying the elite controlled official justice system and by calling for the implementation of Sharia law. When TNSM took over the Swat valley in 2009 it gained some popular support by putting landlords to flight and by encouraging sharecroppers and labourers to cut down their masters' orchards and sell the wood for private profit.[1]

Thus on the one hand Islam was used by governing elites to create national unity and to deflect ethnic and class conflict, but on the other – as events in Swat and Jhang illustrate – outfits wielding its banner were channelling class conflict. So was Islam the ideology of the ruling elites, was it a religion of liberation or was it both? This chapter examines these questions through an examination of lived-Islam in a Punjabi village.

Order, power and prosperity

Islamisation in Pakistan didn't result in the creation of a monolithic Islamic state whose laws, economy and social mores were governed by Islam. In Bek Sagrana, as in much of Pakistan and the Muslim world, people's everyday behaviour fell far short of idealised Islamic standards set by religious scholars and ideologues (see Roy 2007). Not only did the majority of villagers not even pray once a day, but a number of them drank, took drugs, had illicit sexual affairs and even gambled.

Although Islamisation failed to reform people's everyday practices, it did succeed in orienting popular aspirations towards moral and cultural purity, Islamic identity politics and the promise of a highly utopian future – what Bayat (2007) terms 'cheap Islamisation'. In practice this meant that people saw strict adherence to Islam as the panacea for all of Pakistan's social ills and, conversely, people's lack of adherence as the cause of these ills. Thus social issues like poverty, inequality, fraud, and violence were all given religious rather than secular explanations. If these problems arose it was because God was punishing people for straying from the straight path of Islam. But if Pakistanis only adhered to this straight path, God would reward them, their problems would vanish and Pakistan would lead the Muslim world back to its former glory.

In Bek Sagrana and in much of the rural Punjab, the idea that worshipping God could bring peace, prosperity and power was expressed in an Islamic tradition centred on Sufi saints. In this tradition, prosperity and genuine power came from God and was mediated by both orthodox and heterodox saints. Mere worldly power – epitomised for most villagers at the time of fieldwork by economic and political power of the United States – was ultimately sterile. The village Imam told me the following story which clearly illustrated this. During the British Raj, a British soldier had met the

---

[1] The work of Marsden (2007) illustrates similar developments in Swat where madrassa educated scholars challenged the Chitrali nobility in elections in 2002.

pir of Golra Sharif on a railway platform and had asked him what his prayer beads were for. The pir had replied that they were for praying to Allah and in turn asked the soldier what his rifle was for. The soldier told him that it was for shooting and killing, and demonstrated it by shooting and killing a dove on a nearby electric pole. The soldier then taunted the pir to show him if he could do anything comparable with his prayer beads. At which the pir calmly started praying and the dove came back to life. The moral of the story was clear: mere physical force was destructive and ultimately sterile while spiritual power was creative and generative.

In the Punjab as in neighbouring Sindh the descendants of pirs – who were pirs themselves – had come to form a powerful landed aristocracy and the idea that they were channels for God's generative power (*barkat*) was their ideological prop. In this they resembled the Ethiopian and Merina kings described by Donham (1999) and Bloch (1989) whose authority derived from their role mediating divine forces that gave abundant crops, prosperity and blessings to their kingdoms. These notions about fertility and generative power were ideological in the sense that they attributed productivity to a transcendental realm, and in the process, denied the productive role of biological processes and of labour. Thus prosperity, vitality and order were all attributed to pious elders – from the landlord class – who were deemed close to God while the productive role of women, labourers and children who were deemed far from God was denied.

Historically the great Sufi pirs of the Punjab, such as Baba Farid of Pakpattan and Baha-al Haq Zakaria of Multan, established themselves as mediators between local Jat tribes and played a central role in their conversion to Islam. Ansari's (2003) work on the neighbouring province of Sindh indicates that pirs played an important role in settling disputes between tribes and in fostering cooperation between them on issues such as irrigation and trade. Their role as mediators gave them influence and eventually wealth, and as their shrines grew they came to embody both spiritual and worldly power. Although many of the major shrines in both the Punjab and in Sindh were originally founded by ascetic Sufis with little or no material wealth, many of them and their descendants came to form a wealthy landed aristocracy.

These Sufis accumulated their wealth from a combination of gifts from disciples and patronage from the courts in Delhi and later from the British colonial authorities. Through the descendants of pirs – who were often themselves pirs – accumulated vast landholdings and came to form separate, privileged castes. In the area of Pakpattan Sharif the descendants of the 13th century Pir Baba Farid came to form a 'separate "Chishti" caste possessing both economic privileges and ritual status vis-à-vis the local clans' (Eaton 1984: 349). The settlement report of 1892–1899 for the district of

Montgomery reports that the Chishtis owned 9 per cent of all the land in Pakpattan Tehsil (Gilmartin 1984: 224). Richard Eaton reports that this land was acquired by the Chishtis over several generations as a result of gifts (futuh) from disciples, as well as patronage from the Delhi court starting during the Tughluq Period in the 14th century.

Successive rulers in Delhi extended their power in the Punjab through Sufi saints, co-opting them with patronage. Thus the Delhi court used the guardians (*sajjada nishin*) of Baba Farid's shrine to collect revenue. It also helped them build magnificent tombs for their dead ancestors and donated land to them. Eaton argues that as a result of the incorporation of shrines such as Baba Farid's into the larger political structures centred in Delhi, the pirs adopted various symbols, terms and titles from the Indo-Islamic royal courts, including some pertaining to tax collection. He also draws a particular parallel between the dastar bandi ceremony, in which a pir symbolically bestowed legitimate authority by tying a turban on a chief's head, and coronation ceremonies.[2]

The region around Bek Sagrana had itself historically come under the influence of a family of Makhdoom pirs who had been donated land by the numerically dominant Gondal and Ranjha Jats of the region. The land had been donated in exchange for spiritual intercession and mediation in land disputes. An ancestor of the Bek Sagrana Gondals had allegedly donated over 100 acres of land to a Makhdoom pir after his prayers for a son were granted through the pir's intercession. As a result of these and other land grants – namely by the British – the Makhdooms ended up owning over 1,000 acres and forming part of the landed gentry of Sargodha district. In the not so distant past they had apparently kept large numbers of tent

---

[2] After independence, successive governments who saw the social and political influence of the major shrines tried to take control of them redefine their role. Katherine Ewing (1990) describes how all of these governments tried to recast the shrines and pirs light that they felt was congruent with government social and political goals. Thus, General Ayub Khan tried to rationalise the cult of saints by distributing pamphlets on the urs of important pirs that omitted any reference to their miracles and stressed their piety and role as social reformers. In order to further emphasise the secular role of the pirs in promoting social welfare he also built hospitals at various shrines. Zulfikar Ali Bhutto was not as concerned with stripping the cult of saints of superstition as he was with stressing the popular social welfare imparted at the shrines. In a position that combined elements of Khan's and Bhutto's, General Zia emphasised the role of pirs in promoting social welfare policies aimed at a particular type of Islamisation, but returned to a policy designed to strip the cult of saints of superstition and stress the piety and Sharia-mindedness of the pirs.

pegging horses as well as hare coursing dogs and even invited British guests over to go hunting in the Marshlands along the ancient course of the river Chenab behind the village.

The power and wealth of the Makhdooms – like that of the Chishti's at Pakpattan – didn't contradict their spiritual power, it confirmed it. Although as the story recounted above about the saint and the British soldier illustrates spiritual power was superior to worldly power, it didn't exclude it. Spiritual power encompassed and therefore included worldly power. Truly pious men could automatically gain riches and power if they wished but never did so as an end in itself – what happened when they did will be illustrated below. By denying the false idols of money and power and by worshipping God – the single and only true power in the universe according to the Islamic principle of *tauheed* – they acquired powers that were infinitely superior to that of kings, presidents and generals. Many of my informants believed that some pirs were so powerful that they could be in several places at the same time and that they could even destroy and recreate the world in the blink of an eye without anyone even noticing that anything had happened. They believed that these powerful saints secretly governed the universe. One Gondal told me that whatever happened in the world was in fact determined by a powerful hierarchy of saints whose motives were obscure to the uninitiated. When we visited the Chishti shrine at Jalalpur Sharif he told me that the shrine's founding pir was in fact the person responsible for the creation of Pakistan as well as for Pakistan's acquisition of the nuclear bomb. Pakistan's founding father Muhammad Ali Jinnah and Zulfikar Ali Bhutto, the originator of Pakistan's nuclear programme, were merely pawns in his plans to strengthen the Muslim *ummah*.

The idea that the worship of the only one and true God and the denial of worldly idols brought power was vividly illustrated by beliefs surrounding martyrdom (*shahadat*). By dying in Jihad martyrs denied the false idol of the self and thereby affirmed the central tenet of Islam according to which there is only one true God (tauheed). As a result martyrs acquired far greater powers than they ever had in life. Villagers believed that martyrs (*shaheeds*) were alive in their graves and that their bodies didn't undergo biological processes of decay like those of ordinary mortals. The divine favour obtained by martyrs meant that they could intercede spiritually on behalf of supplicants and it was believed that they could secure entry into heaven for 60 people of their choice and 10 others who had learnt to recite the Holy Qur'an by heart. For this reason the graves of individuals martyred in Jihad operations in Kashmir and Afghanistan were often ornately adorned and visited by supplicants seeking intercession.

What pirs shared with martyrs was that they too in a sense died to the world and its false idols. As a result they too served as conduits for divine power (*ruhani takat*) and their shrines were places where people regained their health, their money and their peace. People thought generative power flowed through shrines, much like they thought electricity flowed through electrical grids. The Gondal followers of the unorthodox Pir Alam Shah, frequently invoked Einstein's theory of relativity to explain that God was pure energy. Disciples at this pir's shrine in Sargodha told me when they performed *zikr* at the shrine, it was like turning on a light switch which connected them to God's 'electrical grid'. He went on to compare the power that flowed through this electrical grid to a force greater than 100,000 (*ek lakh*) nuclear explosions.

The generative and fertile aspect of this power was apparent in the notion of *barkat*. The example people most commonly used to illustrate its workings was that if a person cooked for only two people and ten people came to eat there would still be plenty of food to go around for everyone. This could happen on a small scale to ordinary people who lived piously and earned their livelihoods honestly (*halal ki kamai*) but happened on a very large scale at kitchens attached to the shrines of pirs, known as *langars*, where pilgrims and disciples were fed for free (see Werbner 2003). When I visited the shrine at Jalalpur Sharif a disciple of the pir told me that the langar there could never run out of food. He told me that during colonial times the British had told the pir at the shrine to cut down on food expenses at the langar because freely feeding thousands would ruin him. The pir had replied that the langar was no strain whatsoever upon his finances because it was God's endless barkat that flowed through it.

A Gondal pir

In June 2008, one year after his death, Ahmed Abbas Gondal was publicly declared to be a pir by another pir in Sargodha who belonged to the same Sufi brotherhood as he – a branch of the Qadri Sufi brotherhood. The pir also declared that Ahmed Abbas's elder son, Dr Muzaffar Abbas Gondal, was his rightful successor. An upcoming urs in order to celebrate the anniversary of Pir Ahmed Abbas Gondal's death and union with God was then promptly announced. In the days that followed, posters advertising the urs went up on the road leading to the village. The posters proclaimed that Pir Muzaffar Abbas Gondal was holding an urs on behalf of Pir Ahmed Abbas Gondal who had built Astana-e-Alam as a place of worship and had spent over Rs 500,000 on behalf of the poor over the years. His charity and the

Astana were clear signs that Pir Ahmed Abbas was a conduit for God's infinite generative power.

With over 300 acres of land to his name Ahmed Abbas Gondal had been one of the wealthiest landlords in the village of Bek Sagrana. Unlike many of his Gondal relatives he remained aloof from the violent factional politics of the area and devoted most of his life to God. Until his late thirties, and before 1971, he had lived in one of the largest houses in the centre of the village. The two other large adjacent houses belonged to his relatives, the rival factional leaders Ghulam Baksh and Ahmed Rasool. At the time he had briefly worked as a lawyer after taking a law degree at the University of Sargodha. However he quickly renounced the legal profession because it was too corrupt and therefore incompatible with the pious life. Like his father before him Muhamamd Hayat was a devoted disciple (mureed) of Pir Alam Shah of the Qadiri order lineage (silsilah). In 1971, having noted Ahmed Abbas's piety, Pir Alam Shah designated him one of his emissaries (khalifa) and ordered him to build a Sufi lodge, known as an astana, at a designated location at a distance from the village next to the irrigation canal. The astana was to be a place of peace where people could come into contact with God through the practice of remembering God by repeating his various names (zikr) and via Pir Alam Shah who like other pirs mediated between mankind and God. Adjacent to the Astana, Ahmed Abbas was also to build a mosque.

The astana would be a conduit for God's infinite bounty, a place where the poor could obtain free Islamic medication (*hikmati davai*) and food during the monthly celebration of Gyarvi Sharif marking the anniversary of Sheikh Abdul Qadir Gilani's union with God. Pir Alam Shah told Ahmed Abbas to build it in a pleasant and peaceful place, at a distance from the village where corruption, violence and ignorance (*jahliyya*) reigned. The astana, named Astana-e-Alam after Pir Alam Shah, was an eight-pointed star-shaped building. It was located in a four acre compound with large mango and rubber trees and was next to the running water of a large irrigation canal. The trees and the nearby irrigation canal made Astana a cool, shady place where people could escape the torrid heat of the summer months and find peace (*sukoon*) in the greenery.[3] Ahmed Abbas also built two small mosques within the compound and placed handpumps (nalkas) next to them to allow people to carry out their ablutions and have water to drink. It was meant to be a place where wayfarers could stop on their way

---

[3] Green was the favourite colour of the Holy Prophet and it was thought to possess certain curative qualities including that of improving people's eyesight and making people that set their eyes upon it feel at peace.

to pray, but also to rest in the cool shade and quench their thirst. It was a place, much like the Sufi lodge described by Werbner (2003), meant to stand beyond the conflict-ridden, corrupt world that surrounded it, where people could become both physically and spiritually whole.

From 1971 to his death in 2008 Sufi Ahmed Abbas lived as a recluse and spent most of his days within the walls of the compound of Astana-e-Alam. People described him as a fakeer admi, a man who had renounced worldly pursuits for a life of poverty. To him the pursuit of wealth and power was nothing less than idolatry (shirk) because it implied attributing value and agency to things other than God and therefore contravened the Islamic tenet of the unity of God (tauheed). Unlike his ostentatious relatives Sufi Ahmed Abbas led an austere lifestyle. He ate little and rarely consumed meat. He wore simple, frequently worn clothes, and his home was sparsely furnished. He even carried out menial tasks generally reserved for kammi women such as sweeping the ground within the compound of Astana-e-Alam. He never bought a car even though he could have and relied on relatives to drive him and his wife to town. Moreover Sufi Sahib and his sons took pride in the fact that all his earnings were licit (halal ki kamai). Unlike many other Gondals, he didn't steal water to irrigate his orchards nor electricity to run his ceiling fans and water-coolers during the summer.

Over the years several kammi families who at one time or another worked for Sufi Sahib under seyp contracts or as servants and labourers built 17 mud houses around the walls of the Astana-e-Alam compound. The village Imam also moved in and erected the only other brick and mortar house at Astana-e-Alam other than Ahmed Abbas's. Astana-e-Alam also acquired its own shop. In this manner the place became a self-contained unit set apart from the village. This allowed Ahmed Abbas and his family to separate themselves from the factional conflicts that were raging in the village.[4] By virtue of his impartiality Ahmed Abbas acted as a mediator in some of

---

[4] Ahmed Abbas's desire to stay away from the corrupting influence of politics and from the factional conflicts that it inevitably brought was expressed by the fact that throughout his life he sought to remain impartial in the conflict between the Ghulam Baksh Ke and the Ahmed Rasool Ke factions. To this end he went as far as marrying his children into both factions: he had one of his daughters and one of his sons marry with the children of Hajji Sahib (the son of Ghulam Baksh), and he had another son marry the daughter of Ahmed Rasool. On the basis of the impartiality this provided him, during the local council elections described in the previous chapter, for example, Ahmed Abbas instructed all of the kammis living at Astana either not to vote at all or to divide their votes equally between both factions.

the disputes between Chowdri Abdullah and Chowdri Haq Nawaz. I was told that Ahmed Abbas had frequently called members of both factions, including Chowdri Abdullah and Chowdri Haq Nawaz, to the astana and had asked them to settle their disputes at the time when factional conflict in the village was at its peak in the early 1990s.

Elders, authority and spiritual power

Although Sufi Ahmed Abbas was the most prominent pious elder in the area, he wasn't the only one since landlords tended to become pious in their old age. Like Sufi Ahmed Abbas they had adopted a detached attitude towards wealth and power. By renouncing worldly attachments and by worshipping God, they were the ones supposed to sustain the village moral order, and hence ensure people's spiritual as well as material well-being. What happened when they chose worldly goods over God is the subject of what follows.

The term '*bazurg*' used to refer to elders was often used interchangeably with the term pir – both terms derived from Farsi for a wise elder. Like pirs and fakeers, they were supposed to abandon worldly ambition and youthful habits such as drinking and womanising to lead frugal and pious lives. When Chowdris grew old they were expected to spend money on meritorious religious causes such as building mosques and hiring a Maulvi to recite and teach village children the Holy Qur'an. Expenditure on religious causes also frequently included feeding villagers on the monthly celebration of Gyarvi Sharif in commemoration of the birth/death of Hazrat Ghaus Pak. Therefore like pirs who freely fed disciples and supplicants through their langar elders became channels for the bounty and fertility (barkat) of God through their generosity.

As was the case for the Saints in Swat described by Fredrick Barth, elders and pirs were supposed to set aside the aggressive virility of their youth to become gentle to the point of almost being feminine. Chowdris who had hunted (*shikar karna*) or done tent pegging (*neza bazi*) throughout their lives, for example, sometimes publicly renounced these activities when they reached a certain age. These activities were associated with lordly lifestyles characterised by pleasure and dominance. More importantly they were associated with pride and with people's false attribution of power and importance to themselves. This could be interpreted as *shirk*; the violation of the principle of tauheed whereby power ultimately belonged to God and to no one else. By publicly renouncing such activities Chowdris therefore announced that they surrendered to God and renounced egotistical pretence. Similarly it was generally the case that as Chowdris grew old they either grew beards or trimmed their moustaches.⁵

Ahmed Ali, the former head of the Lambar Ke faction of the village, was widely recalled for the fact that he had given up horses and hunting whilst still in his prime and had subsequently lived a simple life. He had influential friends all around the Punjab and was often invited to fairs where tent pegging tournaments were held as well as pig sticking events and hunting expeditions. He had apparently kept a number of stallions tent pegging and went pig sticking. Suddenly, however, at the age of around 55 he renounced both horses and hunting. His pir told him that these activities were pointless (*fazool*) and that killing animals for pleasure was a sin (*gunnah*).

His simplicity and generosity, as well as that of his rival Ghulam Ali, were greatly idealised by villagers and contrasted with the self-importance and greed of their descendants. Elderly village kammis described them to me as simple 'fakeer log' who were concerned with the welfare of villagers. They had sat with villagers at the darra on a daily basis, and knew everyone and their relatives by name. They were also praised for having kept order and punished distillers, drinkers, adulterers, and thieves by fining and even evicting them and for enforcing modesty among both men and women. Women were not allowed to walk around the village without adequate veiling, and weren't supposed to detain themselves to talk to unrelated men, and men were made to wear turbans, particularly before the Chowdris.⁶

---

⁵ This is significant because moustaches were seen as signs of virility and a man without one was considered feminine. The fact that elders trimmed them indicates that excessive virility was problematic and was an obstacle in the way of piety. This is confirmed by the fact that large twirled moustaches associated with particularly virile individuals such as toughs and fierce feudal lords were seen, as were tent pegging and hunting, as signs of pride and self-importance. Villagers frequently told me that it was a sin (gunnah) to grow a large twirled moustache. Finally they were expected to drop the loud and harsh language characteristic of virile and forceful jats and become soft spoken and gentle in their use of words.

⁶ Covering one's head was a sign of respect and subordination both towards God and towards individuals who were considered to embody his will. Men covered their heads during their prayers as well as, generally, when they were in the presence of a pir. Similarly it was considered good form (adaab) to cover one's head before a respected elder.

They and other Gondal landlords had also rotated expenses between them for feeding and providing hookahs to villagers and guests at the village darra. This idealised picture of past village leaders contrasted sharply with the present when the village darra lay abandoned and the Chowdris had largely reneged upon their obligation to grant hospitality to people who had business in the village. The Chowdris were now more concerned with their Lexus Land Cruisers and with their suburban homes than with people's welfare.

In their idealisation of the past, the village had been a more peaceful and in certain ways more affluent place. It had been peaceful because the landlords had imposed order by punishing people who caused conflict by spreading immorality and vice. And it had been affluent in the sense that Marshall Sahlins argued that Hunter-Gatherers were affluent because people allegedly had few and easily satisfied wants. Nowadays on the other hand there was scarcity because despite greater material affluence people's wants had become unlimited meaning that they were never satisfied with what they had. This too led to higher levels of strife because people often pursued their acquisitive desires at each other's expense through the use of force and fraud.

Scarcity, disorder and authority

Landlord elders believed that while they were entitled to govern because of their Godliness, kammis needed to be governed because their insubordination to God made them unable to differentiate what was right from what was wrong. While their own power and wealth was the result of divine favour, the kammi's powerlessness and poverty was the result of divine disfavour. In other words, as is so often the case, elites blamed the poor for their own condition. What was distinctive was that Pakistani elites explained this with reference to religious rather than secular causes. While prosperity and peace were rewards for subordination to God, scarcity, strife and misfortune were attributed to man's insubordination to God. Thus most of my informants interpreted the earthquakes and floods that hit Pakistan in the first decade of the 21st century as evidence of divine wrath. More generally, they believed that contemporary Muslims were backwards and lived in war-torn countries because God was angry with them. God's wrath was even claimed to be visible in the alleged fact that Pakistanis had become smaller, weaker and darker-skinned than in previous generations. Only a generation ago people had been tall and strong and had been able to work for an entire day having had a single glass of buffalo milk in the morning.

Similarly therefore, if kammis were poor it was because God wasn't happy with them. Several of my landlord informants claimed that the very term 'kammi' indicated scarcity and others went so far as to claim that the abusive term 'kamina', referring to poor, dirty, cowardly, and stingy people derived from the term kammi. Whereas pious elders were solely sustained by God, and were therefore constructed as subsisting without labour and independently from human exchange relations, kammis depended on their Chowdri patrons not only for work and a space to live but for moral guidance. According to the Chowdris this was because kammis were dominated by their lower, carnal self (*nafs*). As such they ceded to all of their base impulses making them sexually promiscuous, impecunious, untrustworthy and thieving. As a consequence, and unlike their social superiors, they were weak and unfortunate (*manhoos*) people who needed the guidance and charity of their superiors.

Of all the kammis those held to be most dominated by their nafs and therefore the farthest from Godliness and reason were the Mussallis. They descended from recently converted *Chuhras* who were purported to have eaten carrion and the forbidden (haram) flesh of animals such as wild pigs, snakes, turtles, rats, and lizards. One of their social occupations had apparently been to remove 'night-soil'. The 1888 District Gazetteer for Sargodha reports that so long as they ate forbidden foods and removed 'night-soil' Muslims had refused to eat with them. It was only following their conversion to Islam that Muslims accepted to do so. The term Mussalli, like the alternative term Muslim Sheikh, was granted to them to honour them as new converts. Its literal meaning can be roughly translated as 'he who spends his time on the prayer mat'. Similarly the term Muslim Sheikh was meant to honour them by basically calling them 'Muslim saints'. Yet another term given to honour Chuhras who had converted to Islam was Dindaar or 'faithful' although this term was rarely used. In the village only one family was known as Dindaar and as such it was held to belong to a similar yet distinct category to Mussallis. In spite of the fact that the term Mussalli was originally an honorific it was nowadays practically a term of abuse.[7]

The conversion of the Mussallis to Islam was said to be largely nominal. The joke about the Mussalli who quits a job with a Chowdri where his only task had been to pray five times illustrates this. He was so lazy that he even failed to carry out a simple task that would have gained him religious merit.

---

[7] For an ethnographic account of Christianised sweepers in Karachi, see Streefland (1979).

Mussallis were said to lack diligence in their attendance to the Mosque and Chowdris, and other kammis, claimed that many of them didn't even know how to pray. People also said that they kept dogs and allowed them into their houses and near the cooking hearth (*chula*).[8]

They were also allegedly incapable of honest hard work and were described as 'badshah log' because they behaved as if they didn't need to do any. Instead they liked 'easy work' like distilling, thieving or working as gunmen. In addition to being 'easy', being a gunman allowed them to bully people like lords did. I was repeatedly told stories about Mussalli gunmen who used their position to snatch poultry and even goats from poor people. Their desire to live like lords despite lacking the means was also allegedly apparent by the way they spent their money on lavish weddings and on useless (fazool) luxury goods such as expensive clothes, shoes, TVs, radios, and mobile phones. To finance all this they took loans from landlords and ended up as bonded labourers. Pious Gondal elders and the village Imam frequently lectured them about their impecunious behaviour by telling them that wasteful expenditure was sinful in the light of Islam.

Their earnings – unlike those of pirs such as Ahmed Abbas – were sterile and brought misfortune because they were the fruit of forbidden activities (haraam ki kamai). A Mirasi illustrated this belief by telling me that once after stealing something and selling it he had been surprised by how quickly the money had run out. This was proof that haram ki kamai was sterile. The way he described it, the money of theft had the opposite quality to the fertile (*barkatwala*) quality of money acquired through halal means. For others the money derived from haraam activities also brought misfortune. I heard various stories about people who went blind, deaf or who died as a result of haram ki kamai. Ghulaam Rasool Mussalli had been a gunman and a distiller and had seen three out his four sons die in their twenties. The eldest died of a heroin addiction, the second who was also a gunman died in a shootout and the third was shot by the neighbour who suspected him to be having an affair with his wife.

Kammis, but Mussallis in particular, also lacked the reason and self-control that was the hallmark of the 'respectable' Muslim and of authority. Because they allegedly lacked self-control they were highly active sexually and produced large numbers of children whom they didn't send to school and who consequently hung around the village in dirty rags and causing mischief. These children were thought to be particularly wild and mischievous and adults were weary of finding themselves surrounded by them. When shortly after my arrival in the village, a group of children threw

---

[8] Dogs were considered to be highly impure and contact with them was said to defile and to render both persons and objects impure (pleet/napak).

stones at me calling me a kaffir, villagers were unsurprised to learn that the children in question had been Mussalli children. I was told that this was typical behaviour of uncontrolled, crazy (*pagal*) Mussallis.

Like women and children they displayed excesses of emotion in situations of grief as well as in situations of joy. Mussallis, and Mussalli women in particular, were often criticised for beating their chests and for wailing during periods of bereavement. Such behaviour was said to be un-Islamic because it displayed a lack of acceptance of God's will and certain Maulvis were known to reprimand people for it. It was also associated with the Shia rituals during Muharram that were considered to be un-Islamic by many Sunnis. 'Proper' Sunni behaviour required people to be self-possessed and sober both in situations of grief as well as in situations of joy. Kammis and Mussallis also displayed their lack of self-control by indulging with gusto in music, dancing and generalised revelry at weddings. Their weddings were very lively occasions in which men shouted, sang, danced, and fired shots into the air from rusty shotguns.

Inverting the dominant ideology

Allama Iqbal argued that the cult of saints, which he referred to as 'Persian mysticism' (Iqbal 1964: 81), was an obstacle to democracy by virtue of its creating a spiritual aristocracy pretending to claim power and knowledge not accessible to the average Muslim. More recently, the Pakistani writer Khursheed K Aziz (2001) blamed all social evils on the cult of saints, claiming that it leads to quietism, the denial of personal responsibility, feudalism, dictatorship, and the moral corruption of social life. Abdellah Hammoudi has made a similar claim that authoritarianism in Morocco and the Middle East is rooted in a deeply ingrained respect for authority derived from the authoritarian relationship between Sufi masters and their disciples. He argues that large-scale organisations in Morocco such as government bureaucracies and political parties are governed by criteria of personal allegiance and faithfulness that replicate those governing the interactions between masters and their disciples in Sufi brotherhoods.

However these scholars neglect a rich Sufi tradition of questioning and challenging worldly authority and hierarchy, and wrongly assume that the majority of poor and illiterate rural Pakistanis unquestioningly accept their own subordination. In what follows I explore this tradition of challenging worldly authority and ask whether it can legitimately be labelled as counter-hegemonic.[9]

---

[9] I take this question from James Scott (1985) who argues that subordinate classes use dominant ideologies in a counter-hegemonic manner.

To begin with, kammis generally displayed a highly critical attitude towards individual religious figures. I didn't, for example, meet a single kammi from Bek Sagrana who believed that Sufi Ahmed Abbas was a genuine pir. To them Sufi Ahmed Abbas' frugality and reclusiveness showed miserliness and misanthropy rather than spiritual pre-eminence. Sufi Ahmed Abbas closely scrutinised business operations on his farm, and kept an eye on his servant's every move, making sure that none of them shirked. He had a reputation for overworking and underpaying them, and had four bonded labourers who owed him several hundred thousand rupees each. Moreover he often asked kammis who had settled around his compound on land belonging to him to freely provide him with services and labour. During my stay he asked members of a carpenter (tarkhan) household to set up the electrical lighting for an upcoming urs for free; the carpenters complained that in normal circumstances they could have charged Rs 5,000 for this job but that Sufi Ahmed Abbas claimed that because the work was spiritually meritorious they shouldn't ask for money. People also claimed that Sufi Ahmed Abbas was so obsessed about money that he frequently withheld or reduced people's wages because he suspected them of shirking and stealing from him. They claimed that neither Sufi Ahmed Abbas, nor his two sons who lived in Lahore, ever did anything for them. Sufi Ahmed Abbas was believed to be so hard-hearted that once when the child of a local kammi received an electric shock from a faulty plug in his house and needed to go to hospital, he provided no support and it was the village Imam who ended up taking the child to hospital. His descent into dementia during his last two years of life was interpreted by kammis as evidence of divine displeasure and as a clear sign that he wasn't a genuine saint.

Not only did kammis see Sufi Ahmed Abbas as miserly and hard-hearted, but they rejected his and other landlords' claims that they were bad Muslims. They agreed that there were some individuals among them who were involved in a variety of un-Islamic activities but rightly pointed out that this made them no different to the Gondals. In fact they knew that if anyone was to blame for the spread of drugs and crime in the region it was the Gondals, not them. Furthermore, kammis didn't think that they were bad Muslims because they weren't particularly well-schooled in Islamic theology and didn't pray five times a day. Nor did they buy the claim that the Gondals were more spiritually detached from worldly goods than they were. If anyone was generous and self-sacrificing it was they, not the Gondals. If they were prodigal with their money it was because they had faith that God would provide for them, unlike the landlords who hoarded their wealth.

They interpreted their own behaviour not as out of control and excessively emotional (nafsiati) but rather as passionate, generous and self-sacrificing; they were the mirror image of the Gondals. The term that they often used to describe it was 'jezbati', denoting the willingness to sacrifice oneself for loved ones, as well as for the sake of Islam and to defend the Holy Prophet's name. For them the expression of grief during times of bereavement didn't, as the landlords claimed, constitute a rejection of God's will but was rather an assertion of their love for fellow beings. Landlords on the other were hard-hearted and were purely concerned with the superficial, external aspects of the faith such as praying five times a day with ritual clean (pak) clothes. But they didn't have the leisure to pray five times a day, and besides believed that it was no use doing so if your heart was unclean (napak) or if you were doing it simply to show-off.

In line with these beliefs many kammis thought there was nothing wrong with the Shia rituals of self-flagellation (matam) during the Islamic calendar month of Muharram, when people mourned the martyrdom (shahadat) of Hazrat Imam Hussain who had been martyred fighting the tyranny of Yazid. While for the village Imam these rituals were sinful because they indicated a rejection of God's will and because they harmed the body, for kammis they were signs of people's jezbaat, of heartfelt devotion to Hazrat Imam Hussain and of grief for his martyrdom. For the Imam even simply attending these rituals was sinful. However, because many kammis felt sympathy for Imam Hussain because he symbolised the struggle of the oppressed against corrupt tyrannies, they nevertheless attended these ceremonies despite being Sunnis – although they didn't actively participate. They believed that had it not been for Hazrat Imam Hussain, Yazid would have modified the Sharia to suit his own evil purposes. He would, I was told, have permitted alcohol and free sex. Even worse, he would have made it permissible for people to commit incest.

Their objections however didn't centre so much on people's infringements of Sharia law as much as on the hypocrisy of members of a corrupt elite who preached Islam while they themselves drank, frequented prostitutes, stole government money, and protected smugglers and thieves. In fact many of the religious figures that kammis revered had ignored or even gone against the Sharia laws imposed by the authorities. For them truly pious men were often wandering ascetics (malangs and fakeers) who possessed great spiritual riches but no worldly ones, and who were scorned by temporal rulers and their stooges among learned scholars of Islam (ulema). To these the external rituals of purification and worship associated with the Sharia were meaningless. As such they placed all of the emphasis of spiritual life on practices that fostered purity of the heart. As the work of Katherine Ewing (1984) illustrates, Punjabi wandering ascetics (malangs) saw themselves as being in such close proximity to God and with the purity of the sacred that they could not become polluted and so had no need for external rituals of purification. Ewing claims that the direct contact of malangs with God means that even bodily wastes are not polluting to them. Thus, instead of emphasising the need for prayer and the ritually cleansing ablutions associated with it, malangs seek direct communion with God by inducing themselves into states of intoxication with the use of marijuana (charas).

Views such as these were widespread among kammis with whom malangs often had a shared social background. Thus, an elderly Mirasi told me that even a dog, the most impure of creatures, that had served any of the friends of God – including the Prophets, the companions of the Holy Prophet and the saints – would be dressed up as a man on judgement day and sent to heaven. He said that similarly a poor man who didn't perform his prayers because he didn't possess ritually clean (pak) clothes could also go to heaven. Moreover, like the malangs described by Ewing, kammis often placed a great deal of emphasis on inner states of divine intoxication (diwanagi) as the true form of worship. They claimed that such divine intoxication was often perceived as madness by ordinary mortals and religious scholars.

Kammis frequently referred to Pir Shamsuddin of Multan, also known as Shah Shams Tabrez, as an example of how the Ulema and the worldly authorities often scorned and mistook wandering ascetics who spoke the truth (haqiqat) for madmen. Much like the seventh-century Persian mystic Mansur Al-Hallah, Shamsuddin had claimed that when he spoke it was God and not he who spoke, and the Ulema accused him of blasphemy (kufr) and declared him to be an apostate (murtadd) deserving death. As a result he had gone into exile in Multan where he had been further reviled and driven almost to starvation by people refusing him food. According to one version recounted by a carpenter in Bek Sagrana, it was a poor man who gave him all he had to eat which consisted in some raw half-rotten meat. The fact that it was a poor man who had given the meat was significant because for my kammi informant it clearly showed that the poor were among Shamsuddin's supporters. Like them he was reviled, denied the basic necessities of life, and treated with contempt by worldly and religious authorities. The story goes on to describe that the poor man who had given Shamsuddin the meat couldn't cook it for him because he didn't have any means to do so, and no one else was willing to help. Eventually, because people hadn't left him any other choice, Shamsuddin held the meat up towards the sun, which came down out of the sky and roasted it for him. The sun also almost roasted the local inhabitants, who came running to Shamsuddin asking for forgiveness and pleading with him to send the sun back to its place. In this manner the arrogant and the powerful of the land had been humbled and Shamsuddin, somewhat like Hazrat Imam Hussain, stood as a figure representing the struggle of the poor and of truth against worldly authorities, tyranny and hypocrisy (munafiqat).

The carpenter who told me the story said that landlords were like the people who persecuted Shamsuddin; they preached a heartless, superficial form of Islam that served their interests and felt threatened by anyone who challenged them. They told people to pray, but treated them as if they were animals. One Gondal who was a renowned miser even preached Islam to his servants when drunk. Their Islam was the Islam of power and status, not the Islam of detachment, love and brotherhood preached by the likes of Shamsuddin. Despite their outward displays of frugality and simplicity people like Ahmed Abbas and his sons still liked to mingle with the high and mighty of the land. Being pirs gave them extra social status and made them feel entitled to respect. Likewise, he told me, many people grew beards and became religious preachers because it gave them power and status. It made them feel entitled to cut queues and demand seats on buses. People were scared to say anything to them because they feared being accused of disrespecting Islamic learning. It was also well known, he said, that pirs and maulvis used Islam to make money. Pirs, like the Makhdooms of Daulatpur had made money, got gifts of money, cars and even land for their intercession and had become great lords, and on an admittedly smaller scale the village Imam got *chappatis* from villagers in exchange for his services. All of this contrasted sharply with the practices of unpretentious fakeers such as Shamsuddin who had owned and claimed to own nothing more than a prayer mat (mussalla) and a vessel to carry water (*lota*).

The carpenter also identified with Shamsuddin for personal reasons. His now octogenarian grandfather had once before an assembly of villagers claimed to be God. In subsequent years the old man had intermittently repeated this claim. At the time of my fieldwork the old man walked around the village in the manner of wandering ascetics, dressed in a frayed and dirty shalwar kameez carrying prayer beads in one hand and a long walking stick in the other. He was known to care little for human company and seldom spoke. When he did speak he made prophetic pronouncements that included warning villagers about me – the anthropologist who had allegedly come to pave the way for a British invasion of Pakistan.

As had been the case for Shamsuddin most people other than his close kinsmen thought he was mad (pagal) and paid little attention to him. Like in Shamsuddin's case it was the religious authorities that were the most intent upon denouncing his heterodoxy. The village Imam and his relatives, who happened to be his next door neighbours, said that the old blasphemer had gone mad after walking several miles on a hot summer day. The carpenter's grandson often took me aside to tell me not to listen to them. He said that people who were close to God, like his grandfather, uttered statements which the worldly didn't understand. Without directly referring to his grandfather's claim, he explained that when the great Sufis, such as Shamsuddin, had claimed to be God, it was in fact not they who were speaking but God himself speaking through them. Thus they weren't personally claiming to be God – as the village Imam accused his grandfather of doing – but rather they were claiming to be God's vessels. The village Imam failed to understand this because he was too concerned with money and status. Despite being well off he took bread from the poor on a daily basis, and he liked to hang around with the village landlords.

Conclusion

Clearly kammis didn't accept the idea that their subordinate social position was a result of their moral and spiritual inferiority. This raises the question of whether the ideology that attributed worldly status and riches to proximity with God did actually serve to buttress landlord dominance. I wish to suggest that it did, but not by convincing subordinate classes of their own inferiority. Instead it did so by concealing the social relations that reproduced their subordination. By attributing riches to individual morality and Godliness, the dominant ideology concealed the political and economic inequalities that reproduced landlord dominance. Its result wasn't to silence criticism of the landlords, but to personalise it. In other words, instead of critiquing the class structure, they critiqued the morality of individual landlords. For kammis the problem was that Gondal landlords were immoral, not that they belonged to a structurally dominant class. The corollary to this was that kammis believed that all it would take to redress their problems would be for the Gondals – and leaders more generally – to become genuinely Godly Muslims. As illustrated in the chapter, some elderly kammis believed that what was needed was for the Gondals to be more like their pious and generous ancestors. Others thought that they would be better off if only the Gondals were more like the generous landlords in other villages that they had heard about. On a larger scale, many kammis believed that what they needed was a leader like Zulfikar Ali Bhutto, who many kammis construed as a martyr who was executed under the Zia regime, to give them proper housing, clothes, land, and food.

By situating the source of productivity, prosperity and order in a transcendental sphere beyond human relations, labour and biological processes, the dominant ideology encouraged kammis to believe that worldly advancement would come to them from above rather than through their labour and joint political action. Throughout my stay kammis approached me with requests that I help them find a powerful benefactor to employ them either in Pakistan or abroad. Some even approached me gain access to Chowdri Mazhar Ali. They hoped that a powerful patron would grant them secure housing and employment, help resolve any disputes they might have and help them gain access to medical treatment if ever they or their close kin needed it.

The idea that God Himself could also help directly became apparent to me when a Mirasi domestic servant in the village attempted to convert me to Islam by telling me that if I became a Muslim, God would reward me with riches and happiness. When I told him that there were plenty of Muslims who were neither rich nor happy he replied that it was because they weren't proper Muslims (saccha Mussulman). Those who were proper Muslims might not all be outwardly rich, but all of their needs were satisfied and they were contented. When I then asked him how come there were plenty of rich Gondals who weren't good Muslims, he replied that despite outward riches they were inwardly poor. Unlike genuine Muslims they felt poor because they were greedy and were never satisfied with what they had. Moreover he pointed out that despite their riches they frequently suffered from misfortunes and illness. Sufi Ahmed Abbas had, for example, lost his mind because despite his outward piety he was inwardly more concerned about money than about God. In this way the beliefs associating prosperity with Godliness were unfalsifiable and persisted even though there were plenty of poor people who were selfless Muslims and rich ones who weren't particularly selfless.

CONCLUSION

The 2013 parliamentary elections and their aftermath marked the first ever transition in Pakistan's history from one elected government to another. Before they took place, people from different social sectors had been calling for the suspension of parliament and its replacement with an interim technocratic government because of rampant corruption, crime, terrorist attacks, inflation, and electricity as well as gas shortages. Nevertheless, Asif Ali Zardari's PPP government managed to complete its term in office and to break Pakistan's traditional pattern of corrupt and inept civilian regimes being ousted by generals. But the PPP government achieved more than merely hanging on to power; it also managed to pass major legislation to restore democratic governance and managed to avoid the vendetta-driven winner-takes-all politics of the 1990s democratic interlude. The current PML-N government is now in a stronger position than the PPP was, but it too will face the challenge of consolidating democracy in the face of an interventionist military, an excessively activist judiciary, an unreformed bureaucracy and, last but not least violent militancy and extremism that has seriously undermined the writ of the Pakistani state.

The PPP's most significant achievement was that it managed to cooperate with the opposition PML-N to pass a number of significant amendments to the constitution in order to strengthen democracy. One of the amendments, under the larger 18th amendment, was to rescind General Musharraf's quasi presidential system in which the president could dismiss elected governments. This effectively deprived the military of the tool the military repeatedly used throughout the 1990s to disrupt democratic functioning. Crucially, it also deprived the president of his power to make key appointments to the Electoral Commission of Pakistan and made the process more transparent, consultative and subject to parliamentary scrutiny and approval. The appointment of a Chief Electoral Commissioner now requires consultation between the prime minister and the leader of the opposition as well as by a bipartisan parliamentary committee. The idea behind this amendment was to prevent the Electoral Commission of Pakistan from being an instrument for rigging elections as it was during the local elections of 2005 described in this book. Finally, among the most significant amendments was one reducing the executive's role in appointing superior judges thereby ending the judiciary's subservience to the executive.

While all these measures – and others such as updating electoral registers, making the government hand over power to a caretaker government to oversee elections as a further safeguard against rigging, and making biometric computerised national identity cards compulsory – constituted important steps towards consolidating democracy in Pakistan, an ICG (2013) reports that they weren't sufficient to ensure that the national elections of 2013 were completely free and fair. This is because even though the Electoral Commission of Pakistan wasn't biased – at least blatantly so – it may have been too poorly managed, inadequately resourced, under-staffed, and under-trained to be able to prevent incidents of electoral malpractice. This was the case in 2008, as indicated in chapter five, when rigging near Bek Sagrana actually took place in favour of the PPP rather than the ruling PML-Q because of inadequate police protection of the polling station.

However other reasons why the 2013 elections weren't entirely free and fair included excessive judicial interference during the nominations phase, a disempowered ECP and terror threats. The judiciary selectively and inconsistently applied morality clauses under which candidates could be disqualified if they, their spouses or dependents had outstanding loans or utilities bills and for insufficient knowledge of Islam or for holding views contrary to the 'ideology of Pakistan'. Many were rejected for not being able to recite the call to prayer, to answer certain questions about on Islam. In one instance the respected columnist Ayaz Amir was disqualified for writing two columns that allegedly went against the ideology of Pakistan.

Because the returning officers were members of the notoriously corrupt lower judiciary – rather than from a pool of federal, provincial and local officers selected by the ECP (see ICG 2013) as should have been the case – it is likely that they were subject to what Nelson (2011) calls 'the logic of local politics'; in other words they were subject to pressures or to the influence of local politicians. Chapter five illustrated how this had been the case in 2005 when Chowdri Mazhar Ali Gondal managed, through his contacts in the judiciary, to obtain a returning officer willing to overlook rigging in his favour. If this was possible in 2005 it was surely possible for influential landlords to get sessions judges to reject the nominations of their rivals in 2013.

Moral policing in the name of Islam by an often conservative activist judiciary[1] not only undermined democracy by selectively and arbitrarily preventing candidates from contesting elections but also arguably by even further discouraging the poor from participating as candidates. Few kammis in rural areas like Bek Sagrana are likely to have passed the Islamic knowledge and morality tests of returning officers on the off chance that they wanted to present themselves as parliamentary candidates. As illustrated in chapter seven, poorer villagers' version of the faith didn't stand up to the scrutiny of the orthodox.

Leaving the judiciary aside for the moment, what sort of future might democracy – if the Pakistan military lets it get established – hold for Pakistan? Will politicians, like those described in this book, continue to undermine governance by diverting public resources to their narrow cliques of factional supporters, or will they abandon winner-takes-all politics and dedicate themselves to strengthening state institutions by making them deliver their services along impersonal bureaucratic lines? If politicians follow the first route, then they will continue fuelling political instability. Winner-takes-all politics is a zero sum game in which one faction's political and economic gain is another's loss. Whenever one faction happened to be aligned with the ruling coalition, it distributed state resources to its members and made sure that rivals were deprived of them and even that they were actively victimised and persecuted with the help of the police and even of the secret services to ensure that they remained weak. If politicians start behaving like this again, they will continue fuelling the vendettas that caused so much of Pakistan's political instability in the past – particularly throughout the 1990s when members of the PPP and the PML-N did everything in their power to undermine each other.

This book – and chapter four in particular – illustrated what these vendettas entailed. In Bek Sagrana they meant that members of rival factions were at best trying to deprive each other of things like access to telephone

---

[1] The fact that some members of the judiciary hailed Qadri as a hero for killing Punjab governor Salman Taseer – because he was angry with him for making calls to amend the country's blasphemy laws – is a sign that a judiciary free of military interference won't necessarily result in a more 'liberal' Pakistan in which the minorities can fully participate as citizens.

lines and at worst trying to kill or get each other jailed.[2] However, four argued that the principal casualties of these vendettas were not political leaders themselves but their gunmen as well as ordinary villagers who risked being caught in the crossfire of gun battles. Additionally ordinary villagers were victims of harassment at the hands of the criminals that local politicians patronised to fight each other.

Winner-takes-all patronage politics may also give rise to political instability in another way by fuelling the grievances that feed the various militant organisations that thrive in the country. Lieven (2011) and Lyon (2004) argue that patronage makes life more tolerable for the poorest and therefore prevents large scale popular unrest, but this book showed – chapter three in particular – how its chief effect is to in fact divert state resources towards the rich and away from the poor. Thus one of the central arguments of this book is that the poor in Pakistan are vulnerable precisely because of how patronage politics undermines public sector services. If patronage can in any way be said to prevent large scale popular unrest, it is – again as argued in chapter three – because it keeps subordinate classes subservient and divided rather than because it makes life more tolerable for them. However even here the evidence presented in chapter six indicates that patronage is failing to keep the disaffected poor fully subservient. Thus, in places like Swat and Jhang, the poor are venting their grievances by joining militant outfits that decry the moral corruption and the abuses of the landed elites. One reason for this trend is that although landlords continue to mediate access to the state, their shift away from sharecropping and their move to cities and towns has eroded traditional relations of mutual interdependence – of the type described by Breman (1974) in the context of Gujarat. While they used to live in villages and attended to people's everyday needs from the village men's house, many have moved to town and are no longer readily accessible to their clients.

But what are the chances that electoral democracy will deliver good governance and political stability in Pakistan? One way it could help bring these about is if repeated elections force the landed political elite to become more accountable, or even help replace it with a new political class from a different social sector. The chance that this happens depends on whether elections coupled with economic change will undermine the structures of landlord dominance. This book indicates that despite the erosion of

---

[2] When he got his influence back after Nawaz Sharif rose to power in the Punjab, Chowdri Mazhar Ali managed to get Chowdri Abdullah's younger brother sent to jail for illegally cultivating rice on government land.

traditional ties of interdependence and intermittent elections, the landed elite has retained both political and economic power. Moreover it has done so through the state, much like Indian landed elites.[3]

Evidence from India suggests that even if Pakistan remains a democracy, its traditional elite may not lose power to subordinate social classes. Jan Breman's work illustrates how, thanks to elections, the demographic strength of lower caste Dhodias in a Gujarati village began to be reflected in its leadership structure from the 1980s. However it also illustrates how the Dhodias still lacked the social capital necessary to gain benefits for themselves and their supporters. Additionally it illustrates how whatever the upper caste Anavil Brahmins may have lost in terms of local dominance they regained in terms of contacts and influence at higher bureaucratic and political levels. Similarly work by Jeffrey, Jeffery and Jeffery (2008) illustrates how although Dalit involvement in the politics of Uttar Pradesh may have given them self-esteem and curbed upper-caste atrocities against them, they haven't been substantially empowered. Like Breman's work it shows how local lower-caste politicians lacked the connections in the bureaucracy and with higher level politicians to effectively broker resources for their followers. It also shows how the efforts of lower caste politicians tended to benefit friends and relatives rather than their community as a whole. Finally their work shows how electoral politics co-opted Dalit elites into the political system and therefore undermined the emergence of possibly more radical Dalit politics.

Evidence from India also indicates that increased political competition may sometimes not only fail to undermine traditional elite dominance, but that it may also even lead to increased corruption and criminality. In India increased electoral competition from the 1970s onwards meant that politicians needed to provide more 'clientelistic goods' (see Kitschelt and Wilkinson 2007) – which could range from cash and booze to free bicycles and food handouts – in order to obtain votes. Heightened levels of clientelistic provision meant that political parties and politicians needed substantially more money than previously to even have a chance of winning elections. To obtain it politicians asked bureaucrats to extract money from citizens using their services and pass a percentage of it on to them. They also tapped in to black markets in the liquor, drugs and arms trade. Illustrating these trends, Jeffrey (2000, 2001) has shown how police officers acquire

---

[3] Jeffrey (2000, 2001), Harris-White (2003), Pattenden (2011), and Shah (2010) all show how landlords use their influence within the Indian state to make money and shore up class and caste power against greater low-caste assertiveness, thereby deepening social inequalities.

illegal incomes through bribes, extortion and payoffs, some of which they keep and some of which they pass on upwards through the bureaucracy and up to senior politicians, and Robert Wade (1982; 1985) has illustrated similar processes within the irrigation department.

Finally Paul Brass (1997) has vividly demonstrated how rising levels of electoral competition turned security and safety – principally through control over the use of force and over the police – integral to the struggle for power and influence. The result, he argues, was a Hobbesian state in which security and safety weren't provided by the state. His work and that of Kanchan Chandra shows how it also resulted in communal violence as rival ethnic/caste groups vied against each other to gain control over the spoils of power at each other's expense.

The critics of democracy in Pakistan may conclude that the country would be better off under a benevolent dictatorship – such as General Ayub Khan's and General Musharraf's are often portrayed to have been by the media – than under such gangster politicians. However this would be to ignore the fact that governance under military rule wasn't vastly superior to what it was under democratic rule. Chapter five on devolution and local elections illustrated how even the policies of an authoritarian and technocratic regime could be subverted by the landed class. Despite all of its good intentions and international backing, the Musharraf government was unable to create a new class of accountable local politicians because landlords captured almost all of the local assembly seats. The chapter also illustrated how in some instances Musharraf's military government was even unable to prevent ballot booth capture by powerful opposition politicians.

Pakistan's experience with technocratic government should also serve as a warning to people who believe that the country's problems could be solved if selfless, pious and honest men ruled the country and that Sharia law was properly implemented. As illustrated in chapter six, many Pakistanis believe that the various social evils that afflict Pakistan are due to people's deviation from the straight path of Islam. They believe that the lack of impartial government and justice – to which Islam is committed – is to blame for poor governance, poverty and violence. As illustrated in this book, these are indeed serious issues but it is unclear that honesty and piety alone will allow politicians to bring about social justice in Pakistan. As argued in chapter six, such views ignore the ways in which social inequality undermines governance by preventing the grievances of the poor from being heard. Arguably they ultimately serve the interests of Pakistan's dominant social classes by diverting people's attention away from social inequality.

This book makes it clear that the rule of law and good governance are indeed necessary for equitable social development. Nor do the arguments presented in it in any way contradict claims that honest, law-abiding politicians are a good thing. This may appear to be stating the obvious, but it is important to remember that prominent authors such as Partha Chatterjee (2004, 2011) have argued that the rule of law mainly serves to protect the interests of the propertied classes and tends to further the interests of corporate capital. For example episodes of primitive accumulation, such as the enclosure movement in 19th-century Britain, or its contemporary equivalent in places like central India where the state is handing over the land of indigenous communities to mining companies, have and still do take place under cover of the rule of law.

While Chatterjee believes that civil society and the rule of law curtail popular freedoms by serving corporate interests he believes that it is through the means of 'political society', on the margins and even outside the law, that popular freedoms are being expanded throughout the postcolonial world. In political society, popular pressures force politicians to deliver goods and services through clientelistic exchanges that break and circumvent both the law and bureaucratic procedure and, in the process, make the distant and rigid institutions of the post colonial state more accessible and accountable to the common man. Chatterjee uses the example of slum dwellers who avoid eviction thanks to patrons who create legal and bureaucratic exceptions that allow them to stay where they are. Although these ad hoc arrangements are neither secure nor permanent, Chatterjee believes that they at least prevent the slum dwellers from being evicted as they would if 'bourgeois' state law were implemented.

While it is undoubtedly true that certain laws solely benefit capitalist interests there are plenty of others that, if properly implemented, should benefit the poor. In Pakistan for example it isn't state law that deprives women and sharecroppers of the right to own land but the fact that – as Nelson (2011) demonstrates – it is systematically subverted by politicians who help their predominantly male clients in their efforts to deprive both their womenfolk and their sharecroppers of their legal rights. Similarly the law doesn't deprive people of access to secure housing, landlords who ignore it do. Likewise landlords are the ones who deprive people of health care and education, not impersonal bureaucratic procedures.

This book clearly demonstrates that elites are far less civil and law abiding than Chatterjee seems to allow and that they are often the ones fostering practices that undermine the rule of law and bureaucratic rationality. While it is true that subalterns in Pakistan, as in India, often demand that politicians bend the rules for them, they are not the ones who have the power to ultimately bend and break them. Moreover, as Corbridge et al. (2013) argue, just as Chatterjee's model overestimates the civility of elites it underestimates the civility of subalterns who can often be involved in political movements emphasising institution building and legislation – as illustrated by the case of the grassroots anti-corruption Mazdur Kisan Shakti Sangathan (MKSS) in Rajasthan. I suggest that these rights based movements and solid democratic institutions deserve more credit for the expansion of civil and social rights in India than does political society which – as argued above – often simply reproduces existing inequalities.

By celebrating political society as a potentially richer form of democracy, Chatterjee seems to be equating clientelism with democracy. However genuine democracy involves far more than clientelistic politics; in fact the above discussion indicates that clientelism can undermine democratic institutions, particularly when it operates outside the framework of the law. Genuine democracy involves strong and independent democratic institutions and regular elections but also the freedom to protest, secure civil rights, a lively media, and state acceptance of social movements. If India's subalterns have benefitted from democracy it is thanks to these things and despite the ad hoc and extra-legal nature of political society. It is mainly because 'elections occur regularly, democratic institutions are strong, people are free to protest, and civil rights are guaranteed. The media is lively and social movements are tolerated by the state' (ibid.: 157). Moreover, it is arguably also because directive principles contained in the Indian constitution – including the right to work, to education and to health care – have been upgraded to fundamental rights by supreme court judgments (Birchfield and Corsi 2010: 713). What has in fact impeded democratisation by undermining the effectiveness of public institutions is the ad hoc nature of political society embodied in 'the expansive network of patronage founded on caste and class inequalities' (Corbridge et al. 2013: 157). As in Pakistan, such patronage networks – headed by powerful political brokers – diverted state resources away from the general public along particularistic lines and made people's legal entitlements contingent upon political loyalty.

On the other hand Pakistan has only held elections on an irregular basis and the independence of its institutions has been repeatedly undermined. The judiciary, the electoral commission and almost every other branch of government have all been used to consolidate military regimes and their allies. Moreover social movements have often been dealt with high-handedly. The Anjuman Mazarain Punjab (see Sajjad Akhtar 2006) pressing for tenancy rights on the military farms of Okara has, for example, been severely repressed through brutal police action.

In Pakistan it is arguably the alliance between successive military regimes and the landed class that has perpetuated the arbitrary rule of political society and prevented the emergence of the type of rights-based movement and the consolidation of state institutions that has aided democratisation across the border in India. General Zia-ul-Haq, for example, played an important role in forestalling the emergence of popular movements that could have challenged landed power (see Waseem 1994; Jalal 1995). Moreover, as Hassan Javid (2011) has shown, landed power in Pakistan has largely remained intact because military and authoritarian have repeatedly made alliances with factions within the landed political class. Ultimately, in Pakistan the arbitrary rule of political society can be viewed as the result of this unholy alliance rather than as sign of deepening democratisation. It is this unholy alliance that has perpetuated the inequalities that make the political class unaccountable and perpetuate arbitrary clientelistic politics.

Unfortunately Pakistan's democracy now faces a new and growing obstacle in the shape of violent militant movements. Whether these movements can be defeated will depend on more than mere military force; it will depend on whether Pakistan can achieve inclusive economic growth and greater social justice. But it will also crucially depend on ideological factors and on whether the establishment will continue branding opponents and dissenters as anti-Pakistan and anti-Islam as well as whether it will continue marginalising the poor either by ignoring them or by denying them their social worth because their form of Islam doesn't stand up to their standards of orthodoxy. Last but not least it will depend upon whether women and minorities continue to be denied full citizenship rights in the name of Islam.

BIBLIOGRAPHY

Abou Zahab, Mariam. 2002. 'Sectarianism as a Substitute Identity: Sunnis and Shias in Central and South Punjab', in S. Mumtaz, J.L. Racine and I.A. Ali (eds), *Pakistan: The Contours of State and Society*, pp. 77–95. Oxford: Oxford University Press.

Asian Development Bank. (ADB). 2005. *Improving Devolved Social Service Deliver in NWFP and Punjab*. Islamabad: Asian Development Bank.

———. 2006. *Pakistan Poverty Assessment Update*. Islamabad: Asian Development Bank.

Ahmad, Imtiaz. 1973. 'Introduction', in I. Ahmad (ed), *Caste and Social Stratification among the Muslims*, pp. xvii–xxxiv. New Delhi: Manohar.

Ahmad, Saghir. 1977. *Class and Power in Punjabi Village*. New York and London: Monthly Review Press.

Ahmed, Akbar. 1976. *Millennium and Charisma among Pathans*. London: Routledge and Kegan Paul.

Alavi, Hamza. 1971. 'The Politics of Dependence: A Village in West Punjab', *South Asian Review*, 4(2): 111–28.

———. 1972a. 'Kinship in West Punjab Villages', *Contributions to Indian Sociology*, 6(1): 1–27.

———. 1972b. 'The State in Post-Colonial Societies: Pakistan and Bangladesh', *New Left Review*, 1(74): 59–81.

———. 1973. 'Elite Farmer Strategy and Regional Disparities in the Agricultural Development of Pakistan', *Economic and Political Weekly*, 8(13): 31–39.

Ali, Imran. 1988. *The Punjab under Imperialism 1885–1947*. Princeton, NJ: Princeton University Press.

Ansari, Sarah. 2003[1992]. *Sufi Saints and State Power: The Pirs of Sind, 1843–1947*. Cambridge: Cambridge University Press.

Asad, Talal. 1972. 'Market Model, Class Structure and Consent: A Reconsideration of Swat Political Organisation', *Man*, 7(1): 74–94.

Aziz, Khursheed. K. 2001. *Religion, Land and Politics in Pakistan*. Lahore, Karachi and Islamabad: Vanguard.

Bailey, Fredrick. G. 2001[1969]. *Stratagems and Spoils: A Social Anthropology of Politics*. Oxford: Westview Press.

BIBLIOGRAPHY

Barnett, Steve. A. 1975. 'Approaches to Changes in Caste Ideology in South India', in B. Stein (ed), *Essays on South India*, pp. 149–80. Hawaii: University Press of Hawaii.

Barth, Fredrik. 1959. *Political Leadership among Swat Pathans*. London: Athalone Press.

———. 1960. 'The System of Social Stratification in Swat, North Pakistan', in E. R. Leach (ed), *Aspects of Caste in South India, Ceylon and North-west Pakistan*, pp. 113–46. Cambridge: Cambridge University Press.

———. 1981. 'Segmentary Opposition and the Theory of Games: A Study of Pathan Organisation', in A. Kuper (ed), *Features of Person and Society in Swat: Collected Essays on Pathans*, pp. 55–82. London, Boston and Henley: Routledge & Kegan Paul.

Bayat, Asef. 2007. *Making Islam Democratic: Social Movements and the Post-Islamist Turn*. Stanford, CA: Stanford University Press.

Bearak, B. 2000. Adding to Pakistan's Misery, Million-Plus Heroin Addicts. *New York Times Internet Edition*, 19 April, New York, http://www.nytimes.com.

Bennet-Jones, Owen. 2002. *Pakistan Eye of the Storm*. London and New Haven, CT: Yale University Press.

Birchfield, L. and J. Corsi, 2010. 'Between Starvation and Globalization: Realizing the Right to Food in India', *Michigan Journal of International Law*, 31(1): 691–764.

Brass, Paul. 1965. *Factional Politics in an Indian State: The Congress Party in Uttar Pradesh*. Berkeley: University of California Press.

———. 1984. 'National Power and Local Politics in India: A Twenty-Year Perspective', *Modern Asian Studies*, 18(1): 89–118.

———. 1997. *Theft of an Idol: Text and Context in the Representation of Collective Violence*. Princeton: Princeton University Press.

Brass, Tom. 1999. *Towards a Comparative Political Economy of Unfree Labour*. London: Frank Cass.

Breman, Jan. 1974. *Patronage and Exploitation: Changing Agrarian Relations in South Gujarat, India*. Berkeley: University of California Press.

———. 1993. *Beyond Patronage and Exploitation*. New Delhi: Oxford University Press.

———. 1996. *Footloose Labour: Working in India's Informal Economy*. Cambridge: Cambridge University Press.

Breman, Jan and Kristoffel Lieten. 2002. 'A Pro-Poor Development Project in Rural Pakistan: An Academic Analysis and a Non-Intervention', *Journal of Agrarian Change*, 2(3): 331–55.

Burki, Shahid Javed. 1980. *Pakistan under Bhutto*. New York: Macmillan.

———. 1988. *Pakistan under Bhutto, 1971–1977* (2nd edition), London: Macmillan Press.

Candland, Christopher. 2007. 'Workers' Organisation in Pakistan: Why No Role in Formal Politics?' *Critical Asian Studies*, 39(1): 35–57.

Carstairs, Robert. 1912. *The Little World of an Indian District Officer*. London: Macmillan.

Chakravarti, Anand. 1975. *Contradiction and Change: Emerging Patterns of Authority in a Rajasthan Village*. New Delhi: Oxford University Press.

———. 2001. *Social Power and Everyday Class Relations: Agrarian Transformation in North Bihar*. New Delhi: Sage Publications.

Chandra, Kanchan. 2007. 'Counting Heads: A Theory of Voter and Elite Behaviour in Patronage Democracies', in H. Kitschelt and S. Wilkinson (eds), *Patrons, Clients and Policies: Patterns of Democratic Accountability and Political Competition*. Cambridge: Cambridge University Press.

Chatterjee, Partha. 2004. *The Politics of the Governed: Reflections on Popular Politics in Most of the World*. New York: Columbia University Press.

———. 2011. *Lineages of Political Society: Studies in Postcolonial Democracy*. New York: Columbia University Press.

Chaudhary, Muhammad A. 1999. *Justice in Practice: Legal Ethnography of a Pakistani Punjabi Village*. Karachi: Oxford University Press.

Corbridge, Stuart, Glyn Williams, Manoj Srivastava, and René Véron. 2005. *Seeing the State. Governance and Governmentality in India*. Cambridge: Cambridge University Press.

Corbridge, Stuart, John Harris, and Craig Jeffrey (eds). 2013. *India Today: Economy, Politics & Society*. Cambridge: Polity Press.

Darling, Malcolm. 1934. *Wisdom and Waste in the Punjab Village*. Oxford: Oxford University Press.

Dumont, Louis. 1980[1966]. *Homo Hierarchicus: The Caste System and Its Implications*. Chicago: University of Chicago Press.

Durrani, Tehmina. 1994. *My Feudal Lord*. Reading, PA: Gorgi Books.

Eaton, Richard M. 1984. 'The Political and Religious Authority of the Shrine of Baba Farid', in B. Metcalf (ed), *Moral Conduct and Authority: The Place of Adab in South Asian Islam*, pp. 333–56. Berkeley: University of California Press.

Eglar, Zekiye. 1960. *A Punjabi Village in Pakistan*. New York: Columbia University Press.

Elliot, H. M. and J. Dowson. 1964[1867]. *History of India as Told by Its Own Historians*. Allahabad: Kitab Mahal.

Ewing, Katherine. 1984. '*Malangs* of the Punjab: Intoxication or *Adab* as the Path to God?', in B. Metcalf (ed), *Moral Conduct and Authority: The Place of Adab in South Asian Islam*, pp. 357–71. Berkeley: University of California Press.

Ewing, Katherine. 1990. 'The Politics of Sufism: Redefining the Saints of Pakistan', in A.S. Ahmed (ed), *Pakistan: The Social Sciences Perspective*. Karachi: Oxford University Press.

Fuller, Chris J. (ed). 1996. 'Introduction', in *Caste Today*, pp. 1–30. New Delhi: Oxford University Press.

Fuller, Chris J. and V. Benei (eds). 2001. *The Everyday State and Society in Modern India*. New Delhi: Social Science Press.

Fuller, Chris J. and J. Harris. 2000. 'Introduction', in Chris J. Fuller and V. Benei (eds), *The Everyday State and Society in Modern India*, pp. 1–30. New Delhi: Social Science Press.

Gazdar, H. and H. Bux Mallah. 2011. 'Housing, Marginalisation and Mobility in Pakistan: Residential Security as Social Protection', *CSP Report* Number 4. Karachi: Collective for Social Science Research.

Geertz, Clifford. 1993[1973]. *The Interpretation of Cultures*. London: Fontana Press.

Gellner, Ernest. 1969. *Saints of the Atlas*. London and Chicago: University of Chicago Press.

———. 1977. 'Patrons and Clients', in Ernest Gellner and J. Waterbury (eds), *Patrons and Clients in Mediterranean Societies*. London: Duckworth.

Gilmartin, David. 1984. 'Shrines, Succession, and Sources of Moral Authority', in B. Metcalf (ed), *Moral Conduct and Authority: The Place of Adab in South Asian Islam*, pp. 221–40. Berkeley: University of California Press.

———. 2004. 'Migration and Modernity: The State, the Punjabi Village, and the Settling of the Canal Colonies', in I. Talbot and S. Thandi (eds), *People on the Move: Punjabi Colonial, and Post-colonial Migration*, pp. 3–13. Oxford: Oxford University Press.

Hammoudi, Abdallah. 1997. *Master and Disciple: The Cultural Foundations of Moroccan Authoritarianism*. Chicago and London: University of Chicago Press.

Hanif, Mohammed. 2008. *A Case of Exploding Mangoes*. London: Jonathan Cape.

Hardiman, David. 1982. 'The Indian "Faction": A Political Theory Examined', in R. Guha (ed), *Subaltern Studies I: Writings on South Asian History and Society*. Oxford: Oxford University Press.

Hariss, John. 1980. 'Why Poor People Stay Poor in Rural South India', *Development and Change*, 11(1): 33–64.

Harriss-White, Barbara. 2003. *India Working: Essays on Society and Economy*. Cambridge: Cambridge University Press.

Hassan, A. 2009[2002]. *The Unplanned Revolution: Observations on the Process of Socio-economic Changes in Pakistan*. Karachi: Oxford University Press.

Hassan, Javid. 2012. 'Class, Power and Patronage: The Landed Elite and Politics in Pakistani Punjab'. London School of Economics Doctoral Thesis presented to the department of Sociology.

Herring, Ronald. J. 1983. *Land to the Tiller: The Political Economy of Agrarian Reform in South Asia*. New Haven, CT: Yale University Press.

Hobsbawm, Eric. 2000. *Bandits*. New York: The New Press.

Human Rights Commission of Pakistan (HRCP). 1995. *State of Human Rights in Pakistan 1994*. Lahore: HRCP.

Human Rights Watch. 1995. *Contemporary Forms of Slavery in Pakistan*. London: Human Rights Watch.

Hussain, Akmal. 1989. 'Pakistan: Land Reforms Reconsidered', in H. Alavi and J. Harriss (eds), *Sociology of "Developing Societies" South Asia*. Hong Kong: Macmillan.

———. 2003. *Pakistan National Human Development Report 2003: Poverty, Growth and Governance*. Karachi: UNDP/Oxford University Press.

Ibbeston, Denzil. 1993[1916]. *Panjab Castes*. New Delhi: Low Price Publications.

Inayetullah, M. 1964. *Basic Democracy, District Administration and Development*, Peshawar: Pakistan Academy for Rural Development.

Inbanathan, A. and D. V. Gopalappa. 2002. 'Fixers, Patronage, "Fixing" and Local Governance in Karnataka'. *Working Paper 112*, Institute for Social and Economic Change. Bangalore: ISEC.

International Crisis Group (ICG). 2004a. *Building Judicial Independence in Pakistan*. Islamabad and Brussels: International Crisis Group.

———. 2004b. *Devolution in Pakistan: Reform or Regression?* Islamabad and Brussels: International Crisis Group.

———. 2005a. *Authoritarianism and Political Party Reform in Pakistan*. Islamabad and Brussels: International Crisis Group.

———. 2005b. *Pakistan's Local Polls: Shoring up Military Rule*. Islamabad and Brussels: International Crisis Group.

———. 2005c. *The State of Sectarianism in Pakistan*. Islamabad and Brussels: International Crisis Group.

———. 2011. *Reforming Pakistan's Electoral System*. Islamabad and Brussels: International Crisis Group.

———. 2013. *Parliament's Role in Pakistan's Democratic Transition*. Islamabad and Brussels: International Crisis Group.

International Labour Organisation (ILO). 1993. *World Labour Report 1993*. Geneva: International Labour Organisation.

Iqbal, Muhammad. 1964. 'Islam and Mysticism', in Syed A. Vahid (ed), *Thoughts and Reflections of Iqbal*, pp. 80–86. Lahore: Sh. Muhammad Ashraf.

Jalal, Ayesha. 1995. *Democracy and Authoritarianism in South Asia: A Comparative Historical Perspective*. Lahore: Sang-e-Meel Publications.

———. 1999[1990]. *The State of Martial Rule: The Origins of Pakistan's Political Economy of Defence*. Lahore: Sang-e-Meel Publications.

Javid, Hassan. 2011. 'Class, Power, and Patronage: Landowners and Politics in Punjab', *History and Anthropology*, 22(3): 337–69.

Jeffrey, Craig. 2000. 'Democratisation without Representation? The Power and Political Strategies of a Rural Elite in North India', *Political Geography*, 19(8): 1013–36.

———. 2001. '"A Fist Is Stronger Than Five Fingers": Caste and Dominance in Rural North India', *Transactions of the Institute of British Geographers*, 26(2): 217–36.

Jeffrey, Craig and Jens Lerche. 2000. 'Dimensions of Dominance: Class and State in Uttar Pradesh', in C.J. Fuller and V. Benei (eds), *The Everyday State and Society in Modern India*, pp. 91–112. New Delhi: Social Science Press.

Jeffrey, Craig, Patricia Jeffery and Roger Jeffery. 2008. *Degrees without Freedom? Education, Masculinities, and Unemployment in North India*. Stanford, CA: Stanford University Press.

Jodhka, Surinder S. 1996. 'Interpreting Attached Labour in Contemporary Haryana', *Economic and Political Weekly*, (1996): 1286–87.

Jones, Phillip E. 2003. *The Pakistan People's Party: Rise to Power*. Oxford: Oxford University Press.

Kessinger, T. 1974. *Vilyatpur, 1948–1968: Social and Economic Change in a North Indian Village*. Berkeley: University of California Press.

Khan, M.H. 2006. *Agriculture in Pakistan: Change and Progress 1947–2005*, Lahore: Vanguard.

Khan, M. H. and Dennis R. Maki. 1975. 'Effects of Farm Size on Economic Efficiency: The Case of Pakistan', *American Journal of Agricultural Economics*, 61(1): 64–69.

Kitschelt, I. and S. I. Wilkinson. 2007. *Patrons, Clients and Policies: Patterns of Democratic Accountability and Political Competition*. Cambridge: Cambridge University Press.

LaPorte, R. 2004. 'Implementing Devolution: The New Local Government Scheme', in C. Baxter (ed), *Pakistan on the Brink: Politics, Economics and Society*, pp. 155–70. Lanham, MD: Lexington Books.

Lewis, Oscar. 1958. *Village Life in Northern India: Studies in a Delhi Village*. New York: Random House.

Lieven, Anatol. 2011. *Pakistan: A Hard Country*. London: Allen Lane.

Lindholm, Christopher. 1982. *Generosity and Jealousy: The Swat Pukhtun of Northern Pakistan*. New York: Columbia University Press.

Lyon, Stephen. 2004. *An Anthropological Analysis of Local Politics and Patronage in a Pakistani Village*. Lampeter: Edwin Mellen Press.

Malik, Sohail J. 2005. *Agricultural Growth and Rural Poverty: A Review of the Evidence*, Pakistan Resident Mission Working Paper No.2. Islamabad: Asian Development Bank.

Manor, J. 2004. 'Towel over Armpit: Small-time Political "Fixers" in India's States'. In A. Varshney (ed), *India and the Politics of Developing Countries: Essays in Memory of Myron Wiener*, pp. 60–85. New Delhi: Sage.

Marsden, Magnus. 2005. *Living Islam: Muslim Religious Experience in Pakistan's North West-Frontier*. Cambridge: Cambridge University Press.

———. 2007. 'Islam, Political Authority and Emotion in Northern Pakistan', *Contributions to Indian Sociology*, 41(1): 41–80.

Martin, Nicolas E. 1999. 'The Political Economy of Bonded Labour in the Pakistani Punjab', *Contributions to Indian Sociology*, 43(1): 35–59.

———. 2009. 'The Political Economy of Bonded Labour in the Pakistani Punjab', *Contributions to Indian Sociology* 43(1): 35–59.

Martin, N. 2014. 'The Dark Side of Political Society: Patronage and the Reproduction of Social Inequality', *Journal of Agrarian Change*, 14(3), 419–34.

Mendelsohn, Oliver. 1993. 'The Transformation of Authority in Rural India', *Modern Asian Studies*, 27(4): 805–42.

Michelutti, Lucia. 2008. *The Vernacularisation of Democracy: Politics, Caste and Religion in India*. London: Routledge.

Migdal, Joel. 1988. *Strong Societies and Weak States: State-Society Relations and State Capabilities in the Third World*. Princeton, NJ: Princeton University Press.

Mohmand, Shandana. 2008. 'Local Government Reforms in Pakistan: Strengthening Social Capital or Rolling back the State?' in David Gellner and Krishna Hachhethu (eds), *Local Democracy in South Asia: Microprocesses of Democratisation in Nepal and Its Neighbours*. New Delhi: Sage.

Mueenuddin, Daniyal. 2009. *In Other Rooms, Other Wonders*. London: Bloomsbury.

Mundy, Martha. 1995. *Domestic Government*. London: I. B. Tauris Publishers.

Nasr, Syed V. R. 1994. *The Vanguard of the Islamic Revolution: The Jama'at-I Islami of Pakistan*. Berkeley: University of California Press.

―――. 2000. 'The Rise of Sunni Militancy in Pakistan: The Changing Role of Islamism and the Ulama in Society and Politics', *Modern Asian Studies*, 34(1): 139–80.

―――. 2001. *Islamic Leviathan: Islam and the Making of State Power*. Oxford: Oxford University Press.

Nelson, Matthew J. 2011. *In the Shadow of Shari'ah: Islam, Islamic law, and Democracy in Pakistan*. London: Hurst.

Nicholas, Ralph. 1965. 'Factions: A Comparative Analysis', in K. Banton (ed), *Political Systems and the Distribution of Power*, pp. 21–60. London: Routledge.

Patnaik, U. and Manjari Dingawey. 1985. *Chains of Servitude: Bondage and Slavery in India*. Chennai: Sangam Books.

Pattenden, J, 2011. 'Gatekeeping as Accumulation and Domination: Decentralization and Class Relations in Rural South India', *Journal of Agrarian Change*, 11(2): 164–94.

Pettigrew, Joyce. 1975. *Robber Noblemen: A Study of the Political System of the Sikh Jats*. London: Routledge and Kegan Paul.

Population Census Organisation, Statistics Division, Government of Pakistan, Islamabad. 1999. *1998 District Census Report of Sargodha*. Census Publication No. 36: Islamabad.

Prakash, Gyan. 1990. *Bonded Histories: Genealogies of Bonded Labour in Colonial India*. Cambridge: Cambridge University Press.

Rais, Rasool B. 1985. 'Elections in Pakistan: Is Democracy Winning?' *Asian Affairs: An American Review*, 12(47): 81–102.

Robinson, Marguerite S. 1988. *Local Politics: The Law of the Fishes: Development through Political Change in Medak District, Andhra Pradesh (South India)*. New Delhi: Oxford University Press.

Rose, Horace A. 1911. *A Compendium of the Punjab Customary Law*. Lahore: Civil and Military Gazette Press.

Rouse, Shahnaz. 1983. 'Systematic Injustices and Inequalities: Maliki and Raiya in a Punjab Village', in Hassan Gardezi and Jamil Rashid (eds), *Pakistan, The Roots of Dictatorship: The Political Economy of a Praetorian State*, pp. 311–25. London: Zed Press.

Roy, Oliver. 2007. *Secularism Confronts Islam*. New York: Columbia University Press.

Rudra, Ashok. 1980. 'Local Power and Farm-Level Decision Making', in Meghnad Desai, Susanne H. Rudolph and Ashok Rudra (eds), *Agrarian Power and Agricultural Productivity in South Asia*, pp. 250–80. Berkeley: University of California Press.

―――. 1994. 'Unfree Labour and Indian Agriculture', in Kaushik Basu (ed), *Agrarian Questions*, pp. 75–91. New Delhi: Oxford University Press.

Saif, L. 2010. 'Impact of Colonial Capitalism on the Peasantry in West Punjab', in Birinder Pal Singh (ed), *Punjab Peasantry in Turmoil*, pp. 51–92. New Delhi: Manohar.

Sajjad Akhtar, Aasim. 2006. 'The State as Landlord in Pakistani Punjab', *Journal of Peasant Studies*, 33(3): 479–501.

Salim, Ahmad. 2004. 'Migration, Class Conflict and Change: Profile of a Pakistani Punjabi Village', in I. Talbot and S. Thandi (eds), *People on the Move: Punjabi Colonial, and Post-colonial Migration*, pp. 159–76. Oxford: Oxford University Press.

Sayeed, A. 1996. 'Growth and Mobilisation of the Middle Classes in West Punjab: 1960–1970', in P. Singh and S. Thandi (eds), *Globalisation and the Region: Explorations in Punjabi Identity*, pp. 259–286. London: The Association for Punjab Studies.

Scott, James. 1972. 'The Erosion of Patron-Client Bonds and Social Change in Rural Southeast Asia', *Journal of Asian Studies*, 32(1): 5–37.

———. 1985. *Weapons of the Weak: Everyday Forms of Peasant Resistance*. New Haven, CT: Yale University Press.

Shah, A. 2010. In the Shadows of the State: Indigenous Politics, Environmentalism and Insurgency in Jharkhand, India. Durham, NC: Duke University Press.

Srinivas, M. N. 1987. *The Dominant Caste and Other Essays*. New Delhi: Oxford University Press.

Streefland, Pieter. 1979. *The Sweepers of Slaughterhouse: Conflict and Survival in a Karachi Neighbourhood*. Assen: Van Gorcum.

Syed, Asad. H. 1989. 'Factional Conflict in the Punjab Muslim League, 1947–1955', *Polity*, 22(1): 49–73.

Ullah, Inayat. 1958. 'Caste, Patti and Faction in the Life of a Punjab Village'. *Sociologicus*, 8(2): 170–86.

———. 1963. *Perspectives in the Rural Power Structure in West Pakistan*. Michigan: Development Research and Evaluation Group.

Van den Dungen, P. H. M. 1972. *The Punjab Tradition: Influence and Authority in Nineteenth-century India*. London: George Allen & Unwin Ltd.

Wade, Robert. 1982. 'The System of Administrative and Political Corruption: Canal Irrigation in South India'. *Journal of Development Studies*, 18(3): 287–328.

Wade, Robert. 1985. 'The Market for Public Office: Why the Indian State Is Not Better at Development', *World Development* 13(4): 467–97.

Waseem, Muhammad. 1994. *The 1993 Elections in Pakistan*. Lahore: Vanguard Books.

Washbrook, David. 1976. *The Emergence of Provincial Politics: The Madras Presidency 1870–1920*. Cambridge: Cambridge University Press.

Weiner, Myron. 1967. *Party Building in a New Nation: The Indian National Congress*. Chicago: Chicago University Press.

Werbner, Pnina. 2003. *Pilgrims of Love: The Anthropology of a Global Sufi Cult*. Oxford: Oxford University Press.

Wilder, Andrew R. 1999. *The Pakistani Voter: Electoral Politics and Voting Behaviour in the Punjab*. Karachi: Oxford University Press.

———. 2004. 'Elections 2002: Legitimising the Status Quo', in C. Baxter (ed), *Pakistan on the Brink: Politics, Economics and Society*, pp. 101–30. Lanham, MD: Lexington Books.

Wilkinson, Stephen I. 2011. 'Explaining Changing Patterns of Party-Voter Linkages in India', in Herbert Kitschelt and Steven I. Wilkinson (eds), *Patrons, Cli-*

ents and Policies: Patterns of Democratic Accountability and Political Competition, pp. 110–40. Cambridge: Cambridge University Press.

Wilson, James. 1994[1897]. *Gazetteer of the Shahpur District*. Lahore: Sang-e-Meel Publications.

Wiser, William H. 1936. *The Hindu Jajmani System*. Lucknow: Lucknow Publishing House.

World Bank. 1998. *Pakistan: A Framework for Civil Service Reform in Pakistan*. Islamabad: The World Bank.

———. 2002. *Pakistan Poverty Assessment- Poverty in Pakistan: Vulnerabilities, Social Groups, and Rural Dynamics*. Washington DC: The World Bank.

Yong, Tan T. 2005. *The Garrison State: The Military, Government and Society in Colonial Punjab, 1849–1947*, Lahore: Vanguard.

Zaidi, Akbar. 1999. *Issues in Pakistan's Economy*. Karachi: Oxford University Press.

Zaman, Qasim Z. 2002. *The Ulama in Contemporary Islam: Custodians of Change*. Princeton, NJ, and Oxford: Princeton University Press.

GLOSSARY

ashrafi   People of noble descent, a category of higher caste.
astana   Lodge, abode, threshold.
Barelvi   Name of the religious movement in South Asia originated from the city of Bareilly in Uttar Pradesh, India, whose members venerate Sufi saints and shrines.
begaar   Corvée.
biraderi   Agnatic lineage, in-marrying affinal group, Muslim caste, brotherhood.
chak   Planned village.
charpai   String cot.
chowdri   Honorific term for a landlord, which indicates chieftainship.
darra   Men's house.
dera   A single farm house or a farm house with a small settlement around it.
Eid   Islamic festival. Eid-ul zoha marks the last day of the Hajj, Eid-ul Fitr, to celebrate the end of Ramadan.
fakeer   An ascetic on the Sufi path, mendicant, saint who practices asceticism needy of God's mercy.
ghairat   Shame, decency, honour, control.
goonda   Gangster, muscleman.
Gyarvi Sharif   Communal meal held on the 11th of the month, commemorating the birth/death of Abdul Qadir Gilani of Baghdad.
halal   Lawful, meat of an animal slaughtered in a ritually appropriate way.
haram   Forbidden, taboo, sacred, and protected life.
hayaa   Female modesty.
hookah   Water pipe.
izzat   Honour, respect.
jangal   Jungle, areas outside human habitation.
jangli   Wild person from sparsely populated areas, uneducated, ill-mannered.
Jat   Dominant landowning caste/tribe of the Punjab.

GLOSSARY

Julaha Weaver.
Juma Friday.
kabza Capture, illegal land grabbing.
kaffir Unbeliever, an infidel, a non-Muslim.
kala Black.
kammi Traditional artisanal and menial occupational groups.
khu Well.
Kumhar Potter.
lakh 100,000.
lambardar Generally a landowner who acts as an intermediary between the village and the state in terms of revenue collection and governance.
langar Place where food is cooked and distributed freely at a Sufi lodge in South Asia.
Lohar Blacksmith.
lota Jar with a curved spout that is used for ablutions and for cleaning oneself after going to the toilet. The term is also used to refer to people who frequently shift their political allegiances.
Machi Breadmaker.
Makhdoom Head of a shrine, descendant of a saint.
malang Wandering Sufi ascetic, renouncer, mendicant.
malik Literally the owner of an estate. Also refers to status as a chief.
masjid Mosque.
maulvi A Muslim religious expert on legal matters, cleric.
maund A measure that is equivalent to forty kilograms.
Mirasi Bard.
Mochi Cobbler.
mullah Religious prayer leader or dignitary.
munshi Manager of a landlord.
mureed Lover and desirer (of God), disciple of a Sufi saint.
Mussalli Sweeper.
nafs Carnal, vital soul or spirit, seat of desires.
Nai Barber.
nalka Handpump for extracting water from the ground.
napak Impure.
nazim Elected representative at the union, tehsil or district level.
pagal Mad.
pak Pure.
patwari Revenue officer in charge of keeping land records.
pir Saint, spiritual guide, founder or head of a religious order.
pir-bhai Fellow initiate or spiritual brother of the same saint.
purdah Veil, curtain, female seclusion.

qaum People, nation, tribe. The term can also be used to refer to endogamous lineages and occupational groups.
rassagir 'Holder of the rope', patron who protects thieves (buffalo thieves in particular).
sajjadda nishin Successors, those who sit on the seat of a departed sheikh or saint.
savaab Merit, religious merit.
seyp An informal contract in which village artisans and menials provide services to zamindars in exchange for payments in kind.
Sharia Body of Muslim jurisprudential law, including the Qur'an and Hadith, along with later interpretations.
sharif Noble, good, of noble descent.
silsilah A spiritual genealogy.
takat Strength.
Tarkhan Carpenter.
tehsil An administrative division smaller than the district and larger than the union council.
thana Local police station.
thanedar Local police officer.
urs Religious ceremony held annually to commemorate the death/unification with God of a Muslim saint.
zamindar Landowner.
zat Species, caste, lineage.
zikr Remembrance of God's names through repetition done either alone or in company in order to purify the heart.

INDEX

Abbas, Ahmed 153–6, 160, 165
Abbas, Mahmood 8–10, 12, 14, 56, 130–1
Abbas, Pir Ahmed 153
Abbas, SufiAhmed 10–11, 34–5, 53, 56–7, 60, 155–6, 162, 167
absentee landlords 1–2
agnates 95–7, 116
agriculture 17, 21, 23, 33, 45–6, 50, 72, 76, 88, 124
alliances, instrumental 108
Anavil Brahmin landlords 38
astana 154–6
Astana-e-Alam 153–5
authoritarianism 68

Baksh, Ghulam 98–9, 101, 108, 111, 118, 131, 155
basic health unit (BHU) 85–6, 88–9, 92
begaar 72, 74–5
Bek Sagrana: competition in 97; devolution in 127; factional leadership in 97
Bek Sagrana Gondals 118, 151
BHU see Basic Health Unit
Bhutto, Zulfikar Ali 66, 72, 74–6, 99, 123, 146, 151–2, 166
biraderi 4, 18, 26, 32, 50, 85, 93–6, 103, 116, 118; extended 94–5, 117, 119
birderi 50
bondage 44–65
bonded labourers 11, 45–7, 58, 64, 160, 162
Brass, Paul 173

Breman, Jan 38, 45, 47, 171
bribes 36, 61, 121, 173
brokers, powerful political 143, 175
brotherhood 50

canal colonisation 21, 23–4, 41
CCBs see Citizen Community Boards
central Punjab 20–1
charpais 2–3, 10–11, 29, 34–5, 101
child servants 11, 52, 54–7, 87
Chowdri Abdullah Gondal 9, 26, 84, 98, 100
Chowdri Arif 11, 55–6, 89
Chowdri Asadullah 112–13, 115
Chowdri Mazhar Ali 8–9, 15, 55, 84–6, 88–9, 97–100, 102, 104, 108–16, 118, 127, 130, 166, 169, 171
Chowdri Nawaz Ali 8–9, 11, 14–15, 83–4, 86, 98–100, 103–4, 108, 112, 114–15, 117–18, 127–30, 132, 139–40, 143
Chowdri Rafiq 101–2, 107, 112
chowdris 38, 48–54, 57–62, 96–7, 101, 103–5, 110, 129, 134–5, 142–3, 156–60
Chowdri Sahib 5–6
Citizen Community Boards (CCBs) 124, 129
citrus, harvest 21, 31, 33, 48–9, 52
citrus orchards, introduction of 22–3, 37–8, 46
clientelism 7, 17, 67, 82, 172–6
conflict, factional 116, 155–6
contractors 33, 48–9, 85
contracts 33, 85, 100

189

INDEX

Daulatpur, union council of 127, 130, 132–3, 139–40
debt-bondage 3, 44–6
debts 17, 29, 44–65
democratisation 18, 122, 125, 127, 144, 176
deras 8, 11, 28, 34–5, 37, 40–1, 56, 101, 104, 106, 113, 135, 139
devolution 18, 79, 120, 122, 125–7, 129, 144; in Bek Sagrana 127; elections and 120–44
disciples 19, 150–1, 153, 156, 161
district administration 121–4
district nazim 124–5, 127–8, 143
dominance 66, 69; ideology 161, 166

East Pakistan 70, 72, 146
elections: contesting in 71, 133; devolution and 120–44
electoral politics 66–92
electrical grid 36, 56, 153
empowerment, popular 81, 127–8
enemies 37, 51, 93–119
enmity 108

factionalism 93
factional leadership, in Bek Sagrana 97
families 5, 34–5, 46, 50, 52–6, 58–63, 72, 77, 94, 109, 112, 147, 151, 155, 159
Federally Administered Tribal Areas (FATA) 110
feudalism 8–9, 19
feudal landlords 66–7, 83, 147
friends 93–107, 109–19

gatekeeping 82
generative power 150, 153
generosity 107, 156–7
Gondals: of Bek Sagrana 23; biraderi 94, 99, 116, 118; chowdris 37, 48–51, 53, 59, 75, 135; households 33, 36, 41; landlords 11, 17, 28, 31, 35, 41, 74, 88–9, 110, 129, 158, 166; pir 153; politicians 89, 95, 117; powerful 9, 42, 92; wealthiest 33, 35–6
government, central 67, 93, 120, 123–6, 145–6
green revolution 21–2, 32, 37, 66, 72

households 15, 28–31, 33–4, 36–7, 44, 46, 49, 54, 62–3, 96–7, 132–4, 162; Lohar 30–1; kammi 28, 34–5; Nai 30–1; Tarkhan 30–1
HRCP see Human Rights Commission of Pakistan
Human Rights Commission of Pakistan (HRCP) 45
Hussain, Muhammad 58–62, 133

ICG see International Crisis Group
ideology, dominant 161, 166
ILO see International Labour Organisation
India 7, 20, 32, 45, 69, 83, 172, 174–5
inequalities 66–93, 149, 175–6; reproduction of 66–92
In Other Rooms, Other Wonders (Mueenuddin) 1
International Crisis Group (ICG) 80, 120, 122–4, 126, 136–7, 143–4, 148, 169
International Labour Organisation (ILO) 45
Irfan, Muhammad 59–61
Islam 12–14, 59, 88, 145–67, 169, 173, 176
Islamabad 8, 15–16, 50, 63, 85, 120
Islamisation 18–19, 77, 145, 149, 151

Jeffrey, Craig 16, 172
Jeffrey, Patricia 16, 172
Jeffrey, Roger 16, 172

Kala Gondal 24, 26, 98
kammi households 28, 34–5
kammis 28–31, 34, 37–8, 40–1, 49–51, 57, 64–5, 72–3, 75, 87–8, 91–2, 128–9, 133–5, 158–64, 166

labourers 2, 11, 17, 23, 31–2, 38, 44–7, 49–53, 58, 62–4, 87–8, 131, 133–4, 149–50, 155; attached 48; indebted 45, 63–4; landless 46, 72–3, 135
land: disputes 7, 83, 87, 97, 99–100, 151; donated 151; fertile 20–1; grants 68, 151; reforms 27–8, 67, 71–2, 74–6

INDEX

landed classes 2, 6–7, 32, 67–8, 90–1, 173, 175
landed elites 6–7, 17, 22, 46, 81, 83, 121, 129, 144, 148, 171–2
landed interests 70, 73
landed power 7, 17–18, 116, 122, 144, 176
landlords 1–7, 12–13, 16–18, 21–3, 32–4, 37–8, 44–8, 61–6, 69–72, 74–7, 82–3, 107, 109, 117, 147–9; aristocratic 5, 71, 74; households 17, 32; powerful 3, 13, 19, 27, 42, 70
livestock 23, 26, 32, 35–6, 50, 52, 59–60, 62, 88–9, 109
loans 17, 47, 51–3, 57–9, 62–3, 65, 87, 160, 169
local body elections 75–6, 109, 125–6
local government system 122–4, 126, 143
local politics 4–6, 15, 27, 43–4
local power, maintaining 3
Lohar households 30–1

Makhdooms 4–5, 27, 33, 41, 74, 82–3, 85, 98–100, 102–3, 116, 118, 140, 151–2
malangs 163–4
Malkana, Ghulam Ali 111–16, 128–9, 140
martyrs 147, 152–3, 166
Mekan, Khuda Baksh 114, 127
Mid Gondal 111
military 32, 47, 59, 67, 78–9, 81–2, 144, 168, 176
Mueenuddin, Daniyal 1–2, 4
Mussallis 28–9, 31, 54, 58, 63–4, 87, 91, 135, 159–61

Nai households 30–1
Nawaz, Haq 98, 100, 103–10, 113, 117, 130–2, 138–9, 141–3, 156
neo-bondage 63

Pakistan 97, 147, 168; democracy in 67, 173; politics 9, 66–8, 76
Pakistani Punjab 4, 7, 27, 51, 97, 118
Pakistan People's Party (PPP) 72–6, 78–81, 83–4, 111, 114–15, 124, 126, 128, 168–70

Pakistan State Oil (PSO) 84–5
patronage 79, 171
patterns of settlement and social life 34
pirs 3–4, 26, 73–4, 87, 148, 150–4, 156–7, 160, 162, 165
police officers 9, 15, 100–1, 139, 142, 172
political alliances 18, 95–6, 116–17
political leaders 74, 80
political opponents 71, 79–80, 125, 127
political opposition 76, 122–4, 127, 146
political power 6–7, 32–3, 38, 66–7, 77, 115, 118, 143, 148–9
political society 17, 67, 86, 174–5
political system 71, 81–2, 97, 172
politicians, powerful 99, 112, 129, 132
politics: localising 82, 93–4; parochial 68; winner-takes-all 168, 170
polling 136–7, 139–40, 142; stations 112, 125–6, 132, 137, 139–42, 169
power, spiritual 150, 152, 156
power brokers 100, 128
power structures 82; local 81–2, 93, 120
PPP see Pakistan People's Party
prosperity 30, 145–67
PSO see Pakistan State Oil
Punjabi landlords 47
Punjabi Village 27, 149

Rajputs 42, 142–3
Rasool, Ahmed 99–100, 154–5
referendum, presidential 120, 123, 125–6
rigging 78, 126, 136–7, 141, 143–4, 169–70
rivalries 15, 95–6, 98–9
rural notables 68–70, 75, 82

saints 19, 26, 151–2, 156, 162, 164
Sargodha district 4–5, 16, 20–1, 23–4, 29, 45–6, 84, 92, 148, 151
selflessness 145–67
sharecroppers 22–3, 27–8, 31–2, 46, 128, 149, 174
sharecropping 28, 46, 48, 171
Sharif, Nawaz 78–81, 83–4, 98, 104, 111, 171
shrines 26, 150–1, 153

191

INDEX

South Asia 3, 16, 26–7, 75
Sufi Lodge 10–11, 154–5

Tarkhan households 30–1
tenancy 22–3, 72–7, 87, 95, 114, 128, 134
tribes, agricultural 68

union council elections 108, 128–9, 133, 136
union councillors 125, 127–8

village 20–43; crafts 46; darra 38, 40, 42, 99, 158; factional leaders 52, 116; factions 11, 90, 105, 116; Imam 12, 14, 35, 141, 149, 155, 160, 162–3, 165–6; landlords 24, 166; leaders 18, 26, 38, 44, 140, 158; level 97–8; life 94; population 28
villagers 10–12, 26–7, 36–8, 40–2, 47, 50, 53, 55–7, 86–8, 90, 100–1, 103–6, 108–11, 131–2, 157; landless 17, 26, 75, 86; ordinary 7, 12–13, 49, 119, 133, 171
violence 1, 5–6, 76–7, 131, 146, 148–9, 154, 173

wage labour 17, 28, 31, 46, 48, 50–1
WAPDA *see* Water and Power Development Authority
Water and Power Development Authority (WAPDA) 36
wealthiest chowdris 52–3
weddings 29–31, 51, 59, 64, 87, 102, 161
West Pakistan 70, 72, 145
wheat harvests 28–9, 31–3, 49–50, 52
Wilder, Andrew R. 20, 79, 84, 94, 118, 120, 126
women 15, 35–7, 77, 97, 136, 150, 157, 161, 174, 176